Repair and Rehabilitation of Reinforced Concrete Structures:

The State of the Art

Proceedings of the International Seminar, Workshop and Exhibition

Sponsored by
The National Science Foundation (NSF)
Science and Technology for Development (CYTED)

Organized by
NACE International Latin American Region Venezuelan Section
Venezuelan Corrosion Association (ASVENCOR)
The Center for Hemispherical Cooperation (CoHemis), University of Puerto Rico
The Center for Corrosion Studies—Universidad del Zulia, Maracaibo, Venezuela

Approved for publication by the Materials Engineering Division
of the American Society of Civil Engineers

Maracaibo, Venezuela
April 28–May 1, 1997

EDITED BY
WALTER F. SILVA-ARAYA,
OLADIS T. DE RINCÓN AND
LUIS PUMARADA O'NEILL

Published by

ASCE American Society of Civil Engineers

1801 Alexander Bell Drive
Reston, VA 20191-4400

Abstract: These proceedings, *Repair and Rehabilitation of Reinforced Concrete Structures: The State-of-the-Art,* are from an International Seminar, Workshop and Exhibition sponsored by the National Science Foundation (NSF) and the Science and Technology Program (CYTED) held in May 1997 in Maracaibo, Venezuela. The NACE International Latin American Region Venezuelan Section; Venezuelan Corrosion Association (ASVENCOR); the Center for Hemispherical Cooperation (CoHemis), University of Puerto Rico; and the Center for Corrosion Studies–Universidad del Zulia, Maracaibo, Venezuela jointly organized this workshop whose purpose was to present current international knowledge about reinforced concrete structures and to describe future directions and propose joint research projects for repair and rehabilitation of reinforced concrete structures. Eleven of the fifteen papers are contained in these proceedings covering topics in corrosion, service life, and new materials. Four papers cover diverse subjects such as concrete block deterioration, vibration measurements, stainless steel rebar behaviors, and diagnosis and repair procedures resulting from overloads on a concrete parking structure. Summaries of workshop discussions are presented in Spanish and English.

Library of Congress Cataloging-in-Publication Data

Repair and rehabilitation of reinforced concrete structures: the state-or-the-art: proceedings of the international seminar, workshop and exhibition, Maracaibo, Venezuela, April 28-May 1, 1997 / sponsored by the National Science Foundation (NSF), Science and Technology for Development (CYTED): organized by NACE International Latin American Region Venezuelan Section ... [et al.]; edited by Walter F. Silva-Araya, Oladis T. de Rincón, and Luis Pumarada O'Neill.
p.   cm.
"Approved for publication by the Materials Engineering Division of the American Society of Civil Engineers."
Includes bibliographical references and indexes.
ISBN 0-7844-0299-X
1. Reinforced concrete construction–Maintenance and repair–Congresses. 2. Reinforced concrete–Corrosion–Congresses. I. Silva-Araya, Walter F. II. Rincón, Oladis T. de. III. Pumarada-O'Neill, Luis. IV. National Science Foundation (U.S.). V. CYTED (Organization) VI. NACE International. Latin American Region. Venezuelan Section.
TA683.R347 1997   97-31657
624.1'8341'0288--dc21   CIP

FOREWORD

Constructed infrastructure is essential for the development and progress of commerce and industry in all countries. Most of the buildings, highways, bridges, airports, and transit systems are deteriorating at a rate faster than our ability to renovate them. Reinforced concrete is nowadays the material most used in this context. The repair and rehabilitation of reinforced concrete structures involves a complex interaction of existing and new elements. Engineers are confronted with the continuous challenge of developing new methods to repair, replace or rehabilitate the existing structures.

Several difficulties are encountered by researchers and design engineers in this effort. Engineering judgment, intuition, and experience have been used in many cases due to the lack of technical guidelines in a worldwide context. Different approaches are followed for solving similar problems in different countries, with different degrees of success. Still much information is needed to evaluate the performance, response and reliability of existing and rehabilitated structures. Our knowledge of the capacities of rehabilitation schemes and procedures for retrofitting structures must be improved.

Reinforced concrete is the most common construction material in the Americas. Significant changes in design requirements for such structures have been introduced during the last twenty years, motivated by an increasing knowledge of the behavior of materials, problems with older structures, and earthquakes. Active research continues in order to improve lateral load capacity, ductility, anchorage of reinforcement, column shear capacity, and other aspects of reinforced concrete structures. Structural problems in deteriorated systems, including corrosion, aging, lack of durability, hostile environmental conditions, structural deficiencies, and undesirable seismic responses are matters of continuous testing and research.

Advances in materials science have resulted in the development of new or improved polymers, carbon fibers, composites, polyaramic fibers, and other products. These materials are beginning to play an important role in the repair and rehabilitation of deteriorated concrete structures.

The Structures and Building Systems Program of the National Science Foundation, the Science and Technology Program (CYTED in Spanish), the University of Zulia's Center for Corrosion Studies in Maracaibo, Venezuela, and the University of Puerto Rico's Center for Hemispherical Cooperation in Research and Education in Engineering and Applied Science (CoHemis) jointly sponsored an International Seminar, Workshop and Exhibition titled "The State-of-the-Art of the Repair and Rehabilitation of Reinforced Concrete Structures", held in Maracaibo, Venezuela in April 1997. The purposes of the Seminar/Workshop were to put together researchers and engineers from North and South America to address the state-of-the-art, future directions and possible joint research projects in repair and rehabilitation of reinforced concrete structures; as well as, to provide a mechanism of technology transfer and foster international collaboration.

Potential speakers were identified and asked to provide an abstract of their recent work. These information was used to select sixteen experts from the United States, Latin America, and Europe. The topics covered during the Seminar included corrosion of reinforced concrete structures, durability and service life models, and new materials for repair and rehabilitation. To assure that the Seminar/Workshop was a technically rewarding experience, the activity included the following parts:

- Technical presentations during the three-day Seminar.
- Plenary round-tables on the experts fields with participation of the attendees to the Seminar.
- Discussion group meetings to identify topics for international collaboration in research.
- Exhibition of products and services by local and international companies.
- A one-day field-trip to the repair and rehabilitation works of the General Rafael Urdaneta Bridge in Maracaibo. This is one of the longest cable-stayed concrete bridges in the world.

The 150 attendees from 12 countries included engineers, contractors, government officials, professors, students and researchers involved in the development and applications of technologies in the discussed fields.

This volume contains the papers presented at the seminar, the summary of the workshop discussions and research needs, and other information related to the International Seminar and Workshop. These Proceedings include a broad diversity of on-going research activities and experiences in fields related to rehabilitation of reinforced concrete structures from different points of view and in different countries. Each of the papers included has received two positive peer reviews. All papers are eligible for discussion in the Journal, Materials in Civil Engineering, and are eligible for ASCE awards. It is expected that this document will be of benefit to researchers and practicing engineers by serving as a compendium of the state-of-the-art and future trends in repair and rehabilitation of reinforced concrete structures. Moreover, to emphasize the interest of the organizers in promoting international cooperation, the summary of the workshop results is written in both English and Spanish.

Many people worked hard to make this Seminar/Workshop a successful activity and also contributed to the preparation of the proceedings. The Advisory Committee formed by Dr. Houssam A. Toutanji, Dr. Alberto Sagüés and Prof. Oladis de Rincón were essential during the organization of the technical sessions; also, the invited speakers were available for the peer-reviewing process of the selected papers. Special thanks and recognition is given to the personnel of the Center for Corrosion Studies of the University of Zulia, including Eng. Matilde de Romero, Eng. Miguel Sánchez, Reinaldo González and all their staff and students. Also, special recognition is given to Luz Leyda Vega, Idalia Tomassini and the rest of the staff of CoHemis Center for their assistance since the beginning of this project until the final version of the proceedings.

Walter F. Silva-Araya   **Oladis T. de Rincón**   **Luis Pumarada O'Neill**
Co-Chairman   **Co-Chairman**   **Co-Chairman**

TABLE OF CONTENTS

Foreword
Contents

Workshop

Workshop Program
Participant Information
Breakout Group Topics and Leaders
Breakout Group Participants
Workshop Summary (English Version)
Workshop Summary (Spanish Version)

Papers

Seminar/Workshop Program
"The State-of- the-Art of the Repair and Rehabilitation
of Reinforced Concrete Structures"
April 28-30, 1997

Monday, April 28
(Seminar 9:00 a.m.
-12:35 p.m.)

Coordinators: Miguel Sánchez and Nicolás Nouel

9:00-9:40 a.m.

Oladis T. de Rincón, Universidad del Zulia
*"General Rafael Urdaneta Bridge: Rehabilitation
Based on an Electrochemical Diagnosis"*

9:40-10:20 a.m.

Juan José Carpio, Programa de Corrosión del Golfo
de México. *"Rehabilitation of a Damaged
Reinforced Concrete Bridge in a Marine
Environment"*

10:35-11:15 a.m.

Mario Paparoni, Universidad Metropolitana
*"Structural Damages Caused by Hidden Overloads
in a Reinforced and Prestressed Concrete Car
Parking Building: Diagnostic and Repair
Procedures Performed"*

11:15-11:55 a.m.

Carlos A. Prato, Universidad Nacional de Córdoba
*"Application of Ambient Vibration Measurements
for Repair of the CHACO-CORRIENTES Cable
Stayed Bridge in Argentina"*

11:55 a.m. - 12:35 p.m.

Antonio Nanni, The Pennsylvania State University
*"Strengthening of RC Flexural Members with FRP
Composites"*

2:00-2:40 p.m.

Hamid Saadatmanesh, University of Arizona
*"Durability of FRP Rebars and Tendons in
Aggressive Environments"*

2:40-5:00 p.m.

Workshop
Coordinators: Matilde F. de Romero and
    Alberto Sagüés
Panel Members: Oladis T. de Rincón, Juan J.
    Carpio, Mario Paparoni, Carlos A. Prato,
    Antonio Nanni, Hamid Saadatmanesh

Tuesday, April 29
(Seminar 9:00 a.m.
-12:35 p.m.)

Coordinators: Miguel Sánchez and Nicolás Nouel

9:00-9:40 a.m.

P. N. Balaguru, Rutgers University
"Uses of Inorganic Polymer-Fiber Composites for Repair and Rehabilitation of Infrastructures"

9:40-10:20 a.m.

Houssam Toutanji, University of Puerto Rico, Mayaguez Campus. *"New Approach for Tensile Testing of FRP Composites"*

10:35-11:15 a.m.

Iván R. Lasa, Florida Department of Transportation *"Practical Application of Cathodic Protection Systems for Reinforcing Steel Substructures in Marine Environment"*

11:15-11:55 a.m.

Alberto A. Sagüés, University of South Florida *"Advanced Computational Model for Sacrificial Cathodic Protection of Partially Submerged Reinforced Concrete Marine Footers"*

11:55 a.m. - 12:35 p.m.

Enio Pazini, Universidade Fedeeral de Goiás *"Behavior of Surface and Barrier Coatings on Reinforcement in Cement Mortar Containing Chloride"*

2:00-5:00 p.m.

Workshop
Coordinators: Mary Carmen Andrade, Aleida R. de Carruyo
Panel Members: P. N. Balaguru, Houssam Toutanji, Iván R. Lasa, Alberto A. Sagüés, Enio Pazini

Wednesday, April 30
(Seminar 9:00 a.m.
-12:35 p.m.)

Coordinators: Sebastián Delgado, Mario Paparoni

9:00-9:40 a.m.

Pietro Pedeferri, Departimento di Issica aplicata del Politecnico di Milano. *"Behavior of Stainless Steel Rebar in Concrete"*

9:40-10:20 a.m.

Richard E. Weyers, Virginia Polytechnic Institute and State University. *"Service Life Model for Concrete Structures in Chloride Laden Environments"*

10:35-11:15 a.m.	William Hartt, Florida Atlantic University *"Principles of Condition Assesment and Life-Cicle Planning for Chloride Contaminated Concrete Structures"*
11:15-11:55 a.m.	Mary Carmen Andrade, Institute of Construction Sciences "Eduardo Torroja". *"Calculation of the Residual Life of Postensioned Tendons Affected by Chloride Corrosion Based in Rate Measurements"*
11:55 a.m. - 12:35 p.m.	Silvio Delvasto, Universidad del Valle *"A Case of Deterioration of Concrete Blocks Because of a Zeolite in the Aggregate"*
2:00-3:00 p.m.	Workshop Coordinators: Oladis T. de Rincón, Houssam Toutanji Panel Members: Pietro Pedeferri, Richard E. Weyers, William Hartt, Mary Carmen Andrade, Silvio Delvasto
3:00-6:00 p.m.	Special Workshop Coordinators: Oladis T. de Rincón, Houssam Toutanji, Walter F. Silva Araya

International Seminar, Workshop and Exhibition "The State-of-the-Art of the Repair and Rehabilitation of Reinforced Concrete Structures"

Participant Information

United States and Puerto Rico

Alberto Sagüés   Researcher   (813) 974-5819
University of South Florida   (813) 974-2957 fax
College of Engineering
4202 East Fowler Ave. ENB #118
Tampa, FL 33629

Antonio Nanni   Researcher   (814) 863-2084
The Pennsylvania State University   (814) 863-4789 fax
Department of Architectural Engineering   AXN2@PSU.EDU
212 Engineering "A" Building
University Park, PA 16802

Hamid Saadatmanesh   Researcher   (602) 621-2148
University of Arizona   (602) 621-2266
Department of Civil Engineering   (602) 621-5212 fax
Tucson, Arizona 85721

Houssam Toutanji   Researcher   Toutanji@ce.vah.edu
Dept. of Civil & Environmental Eng.
University of Alabama in Huntsville
Huntsville, AL 35899

Iván R. Lasa   Government   (352) 372-5304
Florida Dept. of Transportation   (352) 334-1649 fax
Corrosion Research Laboratory
2006 N. E. Waldo Road
Gainesville, FL 32609

P. N. Balaguru   Researcher   (908) 445-2232
Rutgers, The State University   (908) 445-0577 fax
Professor of Civil Engineering
P. O. Box 909
Piscataway, NJ 08855-0909

Richard Weyers   Researcher   (540) 231-7408
Professor of Civil Engineering   (540) 231-7532 fax
Virginia Tech
200 Patton Hall
Blacksburg, VA 24061-0105

William Hartt   Researcher   (407) 367-3436
Florida Atlantic University   (561) 367-3885 fax
Post Office Box 3091
Boca Raton, FL 33431-0991

Latin America

Carlos A. Prato   Researcher   (011) 54-51-234262
Universidad Nacional de Córdoba   (011) 54-51-212595 fax
Facultad de Ingeniería   cprato@mafcor.fis.uncor.edu
San Eduardo 151, B. Jardín Espinosa

Enio Pazini   Researcher   (011) 55-62-202-0875
Universidade Fedeeral de Goiás   (Tel./Fax)
Escola de Engenharia Civil
Praca Universitária, s/n
74605-220, Goiania, Brasil

Juan José Carpio   Researcher   (011) 52-981-11760
Programa de Corrosión del Golfo   (011) 52-981-12253 fax
   de México
Av. Agustín Melgar S/N
Apartado 204
24030, Campeche, MEXICO

Mario Paparoni   Researcher   (011) 582-242-5804
Centro de Investigaciones Aplicadas   (011) 582-986-3559 fax
Universidad Metropolitana   mpapar@conicit.ve
Caracas, VENEZUELA

Oladis de Rincón   Researcher   (011) 58-61-598527
Centro de Estudios de Corrosión   oladis@luz.ve
Universidad del Zulia
Maracaibo, Venezuela

Silvio Delvasto   Researcher   (011) 57-2-3392450
Departamento de Ingeniería de   (011) 57-2-3302436 fax
   Materiales
Universidad del Valle
Ciudad Universitaria
Apartado Aéreo 25360
Cali-COLOMBIA

Europe

Mary Carmen Andrade — Researcher — (011) 341-302-0440
Consejo Superior de Investigaciones — (011) 341-302-0700 fax
  Científicas — (011) 341-766-0377 fax
Instituto de Ciencias de la Construcción
  Eduardo Torroja
C/Serrano Galvache S/N
28033 Madrid, Apdo. 19002-28080
Madrid, ESPAÑA

Pietro Pedeferri — Researcher — (011) 39-2-2399-3197
20133 Milano U — (011) 39-2-2399-3180 fax
Vía Mantonelli, 7
ITALIA

International Seminar, Workshop and Exhibition "The State-of-the-Art of the Repair and Rehabilitation of Reinforced Concrete Structures"

A Workshop Sponsored by

National Science Foundation (NSF)
Science and Technology for Development (CYTED)

Organized by

NACE International Latin American Region Venezuelan Section
Venezuelan Corrosion Association (ASVENCOR)
The Center for Hemispherical Cooperation (CoHemis), University of Puerto Rico
The Center for Corrosion Studies-Universidad del Zulia, Maracaibo, Venezuela

Breakout Groups Tracks and Group Leaders

Breakout Group 1
Track: Diagnosis and Inspection of Reinforced Concrete Structures
Group Leader: Oladis T. de Rincón

Breakout Group 2
Track: Evaluation of Repairs and Rehabilitation of Reinforced Concrete Structures with New Materials
Group Leader: Antonio Nanni

Breakout Group 3
Track: Durability Service Life and Long-Term Operation of Structures
Group Leader: Richard Weyers

Breakout Group 4
Track: Corrosion Control Methods
Group Leader: Alberto Sagüés

International Seminar, Workshop and Exhibition "The State-of-the-Art of the Repair and Rehabilitation of Reinforced Concrete Structures"

Breakout Group Participants

Group 1	Group 2	Group 3	Group 4
Oladis de Rincón (Leader)	**Antonio Nanni** (Leader)	**Richard Weyers** (Leader)	**Alberto Sagüés** (Leader)
M. C. Andrade	A. Nanni	H. Toutanji	A. Palacios
P. Castro	A. Rincón	P. N. Balaguru	A. Romero
O. de Rincón	C. Gerbaudo	R. Weyers	A. Sagüés
R. Malavé	H. Castillo	L. Maldonado	E. de Partidas
V. O'Reilly	S. Delgado		E. Pazini
M. Paparoni	S. Delvasto		F. Quintero
E. Ramírez			G. Martínez
A. Sarcos			I. Díaz
			I. R. Lasa
			M. de Romero
			M. Sánchez
			P. Pedeferri
			W. Hartt

INTERNATIONAL SEMINAR WORKSHOP AND EXHIBITION "THE STATE OF THE ART OF THE REPAIR AND REHABILITATION OF REINFORCED CONCRETE STRUCTURES"

Workshop Discussions and Results

Group No. 1: Diagnostic and Inspection of Reinforced Concrete Structures

This group was formed by researchers from Venezuela, Perú, Argentina, Cuba, México and Spain. The group members are actively involved and interested in continuing collaboration in order to reach common objectives. The topics mentioned during the workshop were:

1. *Dynamic techniques for structural evaluation as a supplement to corrosion diagnostic.* Experts in dynamic testing of structures could contribute to supplement the diagnostic of structures with corrosion problems. Technical meetings among structural engineers and corrosion experts are proposed to share experiences and to develop join research projects in this area. Interested countries are Argentina, Venezuela, México and Spain.

2. *Development of criteria for Chloride limits (CL) inducing corrosion on rebars.* Most of the experts in this group have used techniques for detection of CL in diagnostics and prediction of residual life of rebars. Field results show that the criteria for CL limits are not applicable to all situations and circumstances; therefore, there is a need for a better definition of important variables affecting these criteria and to improve predictions. Interested countries are Perú, Spain, México, Cuba, Venezuela, Argentina and the United States.

3. *Degradation of concrete by bacteria and sulfates attack.* Bacteria and sulfates cause serious degradation problems in concrete exposed to industrial environments. There is a need to identify the factors that are causing these problems and develop measures to reduce or prevent their effects. As a first step, the evaluation criteria for both types of degradation must be normalized. Interested countries are Perú, México, Venezuela and Spain.

4. *Revision and discussion of the criteria proposed in the "Manual de la Red DURAR" (Manual of Durar Network) (DURAR=Durability of Rebars, in Spanish).* The manual of DURAR Network (published in Spanish) describes techniques and procedures for evaluation and diagnostic of corrosion in reinforced concrete structures. These methodologies could be applied in the United States and European countries; however, they must be revised and discussed for implementation and general acceptance.

Group No. 2: Evaluation of Repairs and Rehabilitation of Reinforced Concrete Structures with New Materials

New materials are being introduced in the construction industry; however, very few is known about their behavior under different field conditions. Conversely, these materials could be a new alternative for the solution of infrastructure problems in locations where the operating conditions make it very difficult to use conventional materials. For example, the repair of highly contaminated concrete with chlorides and carbonates is extremely difficult and costly under certain conditions, and the durability of the repair cannot be guaranteed. Experimental testing of the behavior of these materials and its possible use in the solution of these problems is necessary. A preliminary plan of action was proposed with the objective of implementing cooperation mechanisms among the interested countries. The United States, Venezuela, Colombia, Argentina, México, Spain, Perú, Italy, Brazil, and

Puerto Rico have shown interest in this initiative. The general steps for the prepared plan are:

1. Select on structural element for evaluation such that it is used in the participating countries. Bridge piers were mentioned as an example.

2. Evaluate the structure under the present conditions by following the procedures outlined in the DURAR Manual in order to validate their applicability.

3. Design the repair considering the local conditions (materials, corrosion, structures, monitoring) and using conventional and new materials.

4. Compare the results obtained with conventional materials and with new materials.

5. Propose recommendations for use of new materials under the studied conditions and possible future research.

Group No. 3: Durability, Service Life and Long-Term Operation of Structures

It is clear that each region has particular necessities for construction materials. High performance materials is a very important requirement in all the countries. Resistance to deterioration as well as some mechanical properties are dependent on the environment surrounding the materials. In the case of concrete, high resistance to contaminants diffusion, particularly chlorides, is very much needed, research topics in this area include:

1. Characterization on the aggressivity of the environment,
2. Feasibility analysis for use and operation of materials under different field conditions and laboratory experiments; both, in rehabilitation and new constructions,
3. Definition of durability criteria,
4. Evaluation and characterization of locally available materials that could be economically feasible.

It is strongly suggested that the private sector be and active participant in this research for the possible commercialization of the news materials developed.

Group No. 4: Corrosion Control Methods

Representatives from public institutions and the private sector are concerned by the lack of existence of a procedure to determine the best technique for prevention of corrosion under given conditions, and the limitations involved. Participants from the United States, Venezuela, Colombia, México, Brazil, Italy and Spain proposed the following research topics in this area.

1. *Behavior of cementicious coating for repair of structures.* The design of concrete mix for the repair of structures is done without proper consideration of technical criteria to achieve a high durability of the concrete. Intuition or experience are the only criteria often used and could be part of the reason why structures deteriorate rapidly. Research must be done in order to establish the factors to be considered in the design of concrete mix for reparation of structures.

2. *Behavior of aminic-alcoholic inhibitors.* Several commercial products are offered as corrosion inhibitors. However there is no information about the characteristics and mechanisms used to control corrosion in steel reinforcement.

Studies must be developed to know the inhibitor, its effect in the mechanical properties of the material, its diffusion capacity and efficiency.

3. *Behavior of galvanized steel embedded in concrete and exposed to chlorides and carbonation.* Galvanized steel is being used as a replacement for steel in environments with chlorides and carbonation; however, there is no information about the conditions and threshold for the beginning of corrosion.

4. *Behavior of stainless steel in the repair of reinforced concrete structures exposed to chlorides and carbonation.* Stainless steel, as a replacement of conventional steel for the repair of reinforced concrete structures could be a viable option and the mixture design would not be a determinant factor.

5. *Comparison of experimental (field) and numerical predictions of throwing power of submerged sacrificial anodes.* Cathodic protection of marine piles could be made at law cost if simple sacrificial anodes submerged in seawater are used. However, the cathodic protection efficiency decays with elevation because of concrete resistance. Computer models exist to predict the maximum elevation for effective protection, but experimental model validation is needed.

Future Actions

A technical workshop with researchers from the United States, Latin America and the European Community was proposed for the preparation of join proposals among the experts interested in the topics mentioned during the workshop session . The main objective of this meeting will be to discuss join research proposal and cooperative projects. Government agencies, industry and universities will be consulted for financial assistance of these collaborative projects. The researchers workshop should be held in Puerto Rico because of its strategic position as a bridge between the United States and the other countries.

The researcher were committed to apply for funding from their own countries, as well as, availability of materials and equipment through local projects to supplement USA funds.

SEMINARIO-TALLER Y EXHIBICIÓN INTERNACIONAL "EL ESTADO-DEL-ARTE EN LA REPARACIÓN Y REHABILITACIÓN DE ESTRUCTURAS DE CONCRETO REFORZADO"

Resultados de las discusiones de las mesas de trabajo

Grupo No. 1: Inspección y diagnóstico de estructuras de concreto reforzado

En este grupo participaron representantes de Venezuela, Perú, Argentina, Cuba y México. Se notó que todos los grupos presentes mantienen investigación activa en los tópicos que se mencionan a continuación y están interesados en continuar impulsando la cooperación para llegar a objetivos comunes.

A continuación se describen brevemente los temas:

1. *Técnicas dinámicas de evaluación estructural como complementarias a las de diagnóstico de corrosión (Argentina, Venezuela, México, España).* En este tema se tienen expertos en pruebas dinámicas de evaluación estructural que pueden verter su experiencia para completar el diagnóstico de una estructura con problemas de corrosión. Se propone inicialmente, tener reuniones técnicas de expertos de estructuras y de corrosión para compartir experiencias que permitan concluir un proyecto de investigación en esta área.

2. *Establecimiento de criterios para límites de CI que inducen a corrosión de la armadura: (Perú, España, México, Cuba, Venezuela, Argentina, Estados Unidos).* La mayoría de los grupos participantes utilizan las técnicas de determinación del CI para el diagnóstico y predicción de vida residual. Los resultados de campo han demostrado que los criterios para los límites de cloruros no son aplicables a cualquier situación o circunstancia, por lo cual es necesario definir las variables más importantes que inciden en estos para establecer criterios que permitan mejorar las predicciones.

3. *Degradación del concreto: Ataque por bacterias y por sulfatos (Perú, México, Venezuela, España).* Se han encontrado problemas de degradación de concreto en medios industriales y marinos debido a ataques de bacterias y por sulfatos. Es necesario identificar los factores que causan estos problemas y proponer medidas para disminuir o eliminar su efecto. Así mismo, y como primer paso, deben unificarse los criterios para evaluar ambos tipos de ataque.

4. *Revisión y discusión de los criterios establecidos en el Manual de la Red "DURAR" (Durabilidad de las Armaduras) (Argentina, Perú, Venezuela, México, España).* El Manual de la Red DURAR (publicado en español) contiene técnicas y procedimientos para evaluar y diagnosticar corrosión en estructuras de concreto armado, que podrían aplicarse en USA y otros países de la Comunidad Económica Europea. Sus resultados pueden ser eventualmente revisados y discutidos para mejorar dicho manual.

Grupo No. 2: Evaluación de reparaciones y rehabilitaciones de estructuras de concreto armado con el uso de nuevos materiales

Al momento, existen materiales de alta tecnología desarrollados para la industria de la construcción, entre los cuales se encuentran algunos cuyo comportamiento no ha sido estudiado en condiciones diferentes de aquellas donde se originó. Por otro lado existe necesidad de resolver problemas en infraestructuras que por su localización y condiciones de operación exigen rehabilitaciones bajo codiciones especiales en las cuales el uso de materiales convencionales no es factible; como por ejemplo la reparación de concretos muy contaminados con cloruros o carbonatados, donde la sustitución de cabillas de acero por otras del mismo material requerirían de una reparación costosa por el volumen de concreto a eliminar que garantice su durabilidad. Esto lleva a pensar en la necesidad de verificar experimentalmente dichos materiales como solución a estas situaciones para dictar recomendaciones respecto a su uso. El grupo discutió un plan de acción preliminar que permita implementar mecanismos de cooperación para trabajar en equipo, con el fin de optimizar los recursos, contando con la participación de aquellos países interesados. Se menciona en primera instancia Estados Unidos, Venezuela, Colombia, Argentina, México, España, Perú, Italia, Brasil y Puerto Rico. Los pasos principales del plan propuesto son:

1. Seleccionar un elemento estructural para ser evaluado y que sea de uso común en los países involucrados. Se menciona en primera instancia los pilotes de puentes con problemas.
2. Realizar la evaluación de la estructura bajo condiciones iniciales siguiendo los procedimientos del Manual de la Red "DURAR" con el fin de validar su aplicabilidad.
3. Realizar el diseño de la reparación tomando en cuenta, para la misma, las condiciones propias de cada lugar (materiales, corrosión, estructura, monitoreo).
4. Comparar las diferencias obtenidas entre el uso de materiales convencionales y el uso de nuevos materiales.
5. Dictar recomendaciones en cuanto al uso de estos materiales en las condiciones estudiadas y posibles investigaciones tendientes a mejorar el conocimiento actual en la materia.

Grupo No. 3: Durabilidad, tiempo de vida útil y operación de largo término

Se ha hecho una breve revisión del estado del arte concluyendo que dependiendo de cada región se requiere tener necesidades específicas. Sin embargo, existe un común denominador que es la necesidad de tener materiales de alta resistencia, no solamente mecánica, sino también a resistir la agresión de los agentes causantes de deterioro y que son intrínsecos al medio en que trabaja el material. En el caso del concreto, se requiere una alta resistencia a la difusión de contaminantes entre los cuales uno de los principales es el ión cloruro. Para una mejor aplicación de los materiales utilizados para construir, reparar o rehabilitar estructuras que tengan larga vida útil es necesario realizar investigación conjunta entre distintos especialistas para:

- Caracterizar la agresividad del medio ambiente.
- Evaluar la factibilidad de operación de los materiales en campo y en el laboratorio tanto para nuevas construcciones como para reparaciones y rehabilitación.
- Definir criterios de durabilidad.
- Evaluar y caracterizar nuevos materiales locales que puedan responder a la necesidad de bajo costo.

Se recomienda involucrar al sector productivo (privado) para hacerlo partícipe de los resultados de las investigaciones y se interese en el uso de los nuevos materiales que resulten de las investigaciones.

Grupo No. 4: Métodos de control de corrosión

En esta mesa de trabajo estuvieron presentes representantes de instituciones públicas y privadas, los cuales manifestaron su preocupación por no existir un procedimiento o técnica que de alguna manera les permitiese utilizar un adecuado método de control, no se da a conocer de una manera clara y convincente la forma de cómo se puede aplicar, y en caso de que se utilice, cuál sería su alcance y sus limitaciones.

Los participantes correspondieron a EE.UU., Venezuela, Colombia, México, Brasil, Italia, España. Entre los tópicos que se discutiera se tiene:

1. *Comportamiento de recubrimientos cementosos en reparación.* Durante reparaciones de estructuras de concreto se utilizan diseños de mezclas, las cuales en la mayoría de las veces no se seleccionan con criterios técnicos para lograr una mayor durabilidad del concreto; sino que algunas veces va a depender de criterios subjetivos, razón por la cual al poco tiempo se deterioran las estructuras. Por eso es importante conocer cuáles son los factores que se deben considerar en el diseño de mezcla a ser utilizada en una reparación.

2. *Comportamiento del inhibidor alcohol amínico.* Actualmente en el mercado se están ofreciendo una serie de sustancias que se están señalando como inhibidores de corrosión, sin embargo, no se indican las características ni los mecanismos bajo los cuales funciona para lograr controlar la corrosión en el acero. Se hace necesario conocer del inhibidor su efecto en las propiedades mecánicas, su poder de difusión y su eficiencia.

3. *Comportamiento del acero galvanizado embebido en concreto y expuesto a Cl- y/o carbonatación.* Se ha venido utilizando desde hace algún tiempo el acero galvanizado como un sustituto del acero en medios con presencia de cloruros

y/o carbonatación, sin embargo, no se han definido las condiciones y umbrales para los cuales se inicia la corrosión.

4. *Comportamiento del acero inoxidable embebido en concreto y expuesto a Cl y/o carbonatación.* La utilización del acero inoxidable en las reparaciones de estructuras de concreto reforzado pudiese ser una alternativa viable, y donde el diseño de mezcla no ser[FML1]ía un factor muy determinante.

5. *Comparación de datos experimentales y predicciones numéricas de "poder impulsivo" de ánodos de sacrificio sumergidos.* La protección de pilástras en ambientes marinos puede llevarse a cabo a bajo costo mediante el uso de ánodos de sacrificio sumergidos, sin embargo, la eficiencia de la protección catódica decae con la elevación debido a la resistencia del concreto. Se dispone de módulos computadorizados para predecir la elevación máxima para protección efectiva, sin embargo, se requiere de la validación experimental de dichos modelos.

Acciones Futuras

Para logar la interacción en todos estos temas se proponen reuniones técnicas con el fin de que los interesados puedan discutir en detalle los mismos y escribir conjuntamente proyectos de cooperación y propuestas que serán sometidas a agencias, industrias y gobiernos que estén dispuestos a financiar proyectos colaborativos. Se proponen estas reuniones en Puerto Rico, por servir de puente entre los Estados Unidos y los demás países.

Los investigadores se comprometen a solicitar en sus propios países apoyo para la compra de equipos y materiales a través de proyectos nacionales.

ADVANCED COMPUTATIONAL MODEL FOR SACRIFICIAL CATHODIC PROTECTION OF PARTIALLY SUBMERGED REINFORCED CONCRETE MARINE FOOTERS

Alberto A. Sagüés*, S.C. Kranc* and Francisco Presuel-Moreno**

ABSTRACT

Calculations to evaluate the extent of cathodic protection that may be achieved in marine substructure footers by means of submerged sacrificial anodes are presented. The case of a half submerged cylindrical footer, 3 m in diameter and 1 m high is examined with a numerical computational model using as sensitivity parameters the concrete electrical resistivity, oxygen diffusivity, and concrete rebar cover. Fully active steel was assumed. The calculations return the corrosion rates of the rebar, as a function of position, before and after connection of a zinc sacrificial anode. Adequate protection of the entire rebar assembly appears to be feasible when the concrete resistivity is low (2 kΩ-cm) and the concrete approaches water saturation. The calculations suggest however that protection above the waterline (and in extreme instances, possibly even below the waterline) is of limited effectiveness for medium to high (10 - 100 kΩ-cm) wet concrete resistivities.

* Professor, Department of Civil and Environmental Engineering, University of South Florida, Tampa FL 33620
** Research Assistant, Permanent Affiliation: CINVESTAV-Mérida México

INTRODUCTION

Sacrificial cathodic protection (CP) of reinforced concrete marine substructures is an attractive method to extend service life after corrosion has developed [Sagüés and Powers, 1996; Kessler et al., 1996]. However, the effectiveness of a CP system is difficult to assess beforehand because of the complicated polarization characteristics of steel in concrete, and of the highly variable resistivity of the medium. An important example of this complexity involves predicting the usefulness of bulk anodes placed in the seawater to protect partially submerged footers. Depending on the value of the concrete resistivity and of the amount of steel and concrete as a function of elevation, the length of the region above the waterline protected by the anodes (the "throwing distance") may only be a few cm, or may adequately cover the entire region of interest. The polarization distribution problem is not amenable to a simple one-dimensional treatment and must take into consideration not only the distribution of electrical potential through the bulk of the concrete, but also that of oxygen. In this paper, an advanced computational model is used to develop exploratory predictions of CP performance in footers.

SYSTEM MODELED

System Dimensions.

The model was applied to a footer shaped as a short cylinder with the lower half submerged in seawater (Figure 1). This geometry approximates the squat prism present in many structures, and permits the reduction of the problem to two dimensions by assuming angular symmetry. The calculations were performed for a footer 3 m in diameter and 1 m high, with the rebar cage placed to result in a uniform concrete cover distance x from the external lateral, top, and bottom footer surface. A single mat of rebar was considered, with a steel placement density such that 1 m^2 of nominal mat area contained 1 m^2 of steel surface in contact with the concrete. This condition approximates typical construction practice. The overall footer dimensions and reinforcing are representative of typical marine construction and although not varied in the cases presented here, the model could reproduce any other dimensions or reinforcing schedule designed. Three different cover thicknesses (x=5 cm, 10 cm and 15 cm) were used in the calculations to evaluate the sensitivity of the results to this construction parameter. The presence of underlying piles or supported columns was not explored in this investigation.

Concrete Properties.

Concrete properties of relevance to the performance of a cathodic protection system include the electrical resistivity ρ and the effective oxygen diffusivity D. For the purposes of these exploratory calculations both properties have been treated as approximately uniform throughout the body of the short footer, which has a less extreme variation in water content than in a vertically extended structure. Because the scale of distribution of current and oxygen flow spans distances much larger than the aggregate size, no local heterogeneities in ρ or D were considered. Three values of ρ were selected for the parameter sensitivity calculations (ρ=2 kΩ-cm; 20 kΩ-cm and 100 kΩ-cm). These values span the range of resistivities encountered in surveys of wet concrete marine substructure elements near the waterline [Sagüés et al., 1994b]. There is little information on effective values of oxygen diffusivity in field structures, but laboratory experiments suggest that concrete nearly saturated with water is expected to have values of D exceeding 10^{-6} cm^2 /sec. (when expressing the oxygen concentration in terms of moles per unit volume of pore water). The value of D increases markedly when saturation is less than complete [Sagüés et al., 1994a; Gjorv et al., 1986, Tuutti, 1982]. Accordingly, values of D=10^{-3} cm^2 /sec, 10^{-4} cm^2 /sec and 10^{-5} cm^2/sec were chosen as parameters for the sensitivity calculations.

Electrochemical Variables.

The calculations treated only the case of a footer that had been in service long enough for severe chloride ion contamination over the entire rebar cage to develop, so that the steel is in the active condition everywhere. Active steel dissolution was modeled by assuming Butler-Volmer kinetic control of the reaction Fe $\rightarrow$ Fe^{++} + 2e with constant activity for reactant and products. Only one cathodic reaction (oxygen reduction, O_2 + 2 H_2O + 4e $\rightarrow$ 4 OH⁻) was considered, also obeying Butler-Volmer kinetics but allowing variation in the activity of oxygen. The kinetic constants for these reactions are given in Table 1. These assumptions are sweeping simplifications of the actual behavior of reinforcing steel in concrete, but represent a reasonable first approximation for the purposes of examining the sensitivity of the cathodic protection scheme to variations in basic properties of the system. Furthermore, the values of the kinetic constants have been chosen to reflect experimentally observed magnitudes and create operating conditions that are representative of conditions often encountered in the field.

The galvanic anode (or anode array) was considered to be large enough to effectively create a fixed potential difference between the rebar cage steel and the seawater immediately in contact with the submerged concrete surface. This condition is reasonable for common commercial anodes since for the cases examined, the total calculated anode currents exceeded 1 A only in two borderline conditions, and were mostly less than 0.1 A. A value of -1V vs SCE was used as a convenient nominal working anode potential representative of commercial Zn alloy products. As the seawater resistivity value ($\rho \approx$ 20 Ω-cm) is much smaller than those assumed for the concrete, the assumption of an equipotential concrete-seawater interface is also a justifiable approximation.

COMPUTATIONAL PROCEDURES

The method used to model the distribution of corrosion has been previously presented [Kranc and Sagüés, 1994] and will be only briefly discussed here. While this modeling technique can examine the effect of spatial variation of properties, for the cases chosen here (concrete with constant ρ and D), the governing equations for the electrical potential E and the oxygen concentration C in the concrete reduce to:

$$\nabla^2 E = 0 \qquad (1)$$

$$\nabla^2 C = 0 \qquad (2)$$

At the outer surface of the concrete boundary conditions are specified in terms of the environment. The oxygen concentration (expressed per unit volume of pore water) is assumed to be a constant at the concrete surface, equal to the concentration in air-saturated water. The surface of the concrete submerged in seawater is assumed to be equipotential. The concrete surface exposed to air is treated as electrically insulated.

At the reinforcing steel-concrete interface, boundary conditions are dictated by the electrochemical reactions indicated earlier. The metal dissolution and oxygen consumption reactions produce currents densities i at the concrete-rebar interface

$$i_a = i_{0a} e^{(E-E_{0a})/\beta_a} \qquad (3)$$

$$i_c = i_{0c}\frac{C}{C_0}e^{(E_{0c}-E)/\beta_c} \qquad (4)$$

for the anodic and cathodic reactions (denoted with subscripts a and c respectively). Appropriate sign conventions were established.

These equations are formulated in terms of local current densities, and are functions of the exchange current density, i_0, the Tafel slope, β, and the equilibrium potential E_0. At the interface, E is the difference of potential between the electrolyte directly in contact with the metal and the metal itself, considered an equipotential surface by virtue of its high electric conductivity. In the interest of computation economy, the rebar cage was envisioned as a pervious sheet so that during computation the exchange current densities are reduced by a factor of ½ from the value shown in

Table 1, to account for both sides of the sheet. After computation, the values of the current densities on the sheet are converted again to the corresponding steel surface magnitudes and reported as such in the Results section.

Applying Ohm's law along the direction of the normal n to the steel surface

$$\frac{\partial E}{\partial n} = \rho \ \Sigma i \qquad (5)$$

Likewise, the equivalent current density due to the consumption of oxygen is given by Fick's first law by

$$\frac{\partial C}{\partial n} = \frac{i_c}{4\,FD} \qquad (6)$$

The factor 4 appears as the number of electrons transferred in the reduction reaction and F is Faraday's constant.

The simultaneous solution to Equations (1-6) is by a Gauss-Seidel procedure, modified to accommodate the nonlinear, implicit boundary conditions at the steel concrete interface. For the freely corroding footer it is necessary to search (using as a parameter the potential difference between steel and seawater) for solutions which minimize the current to the water. For the cathodically protected footer, the potential of the steel with respect to the water is fixed by the working potential of the anode.

The system was modeled by applying finite difference approximations to determine the distribution of electrical potentials and oxygen concentrations. The modeled footer has 80 nodes in the vertical direction and 120 in the radial direction. The grid spacing is 1.25 cm in both directions.

RESULTS

Conditions Before Application of Cathodic Protection

Figure 2a shows the calculated unprotected corrosion current density of the steel (i_u) for intermediate cases in which $D=10^{-4}$ cm^2/sec and x=10 cm while ρ spans the entire chosen range. The values approach those commonly observed at low elevations above the waterline, e.g. corrosion current densities on the order of 1 $\mu A/cm^2$ [Andrade et al., 1994]. Because the calculations were limited to cases with electrochemical parameters that do not vary with elevation, only minimal corrosion macrocell action was obtained. Consequently, the calculated corrosion rates away from the edges were virtually independent of the assumed concrete resistivity. Calculations for other conditions showed that the corrosion rates increased with the assumed values of D or of 1/x (see Table 2), indicating that with the kinetic parameters assumed the cathodic reaction was experiencing various degrees of diffusional polarization. The calculated corrosion rates were greatest at the corners of the rebar cage, which is to be expected since the oxygen supply to that region involves both the lateral and end surfaces of the rebar cage assembly.

The calculated steel potential for the examples in Figure 2a was nearly independent of the value of ρ, approaching - 670 mV SCE against a reference electrode placed in the seawater. This potential is comparable to values commonly reported in the field under similar conditions. Table 3 lists the potential values for other cases, reflecting the overall corrosion rate trends of Table 2.

Conditions Following Application of Galvanic Cathodic Protection

The bottom portion of Figure 2 shows the effect of connecting the galvanic anode. The extent of cathodic protection has been examined by observing directly the changes in the calculated corrosion current density, rather than using an indirect magnitude such as the local variation in potential [Funahashi and Bushman, 1991]. Designating as i_u and i_p the corrosion current densities before and after application of cathodic protection, the figure shows (left ordinate) the current density ratio i_p / i_u, as a function of position in the rebar cage assembly. The right ordinate shows percent protection = 100 (i_u-i_p) / i_u. The calculated protection efficiency was greatest in the low resistivity (2 kΩ-cm) case, in which over 90% reduction in the corrosion rate resulted both above and below the waterline. In the 20 kΩ-cm case protection was efficient below the waterline but decayed to less than 50% about halfway between the waterline and the upper rebar mat. Protection was poor, even below the waterline, for the 100 kΩ-cm case. Clearly, the choice of concrete resistivity is critical in determining both the maximum efficiency and the throwing power of the galvanic system. The results for the 100 kΩ-cm case are particularly interesting since concretes with pozzolanic additions and low w/c ratios (which have wet-condition resistivities that may approach 100 kΩ-cm) are increasingly being used for marine substructures to reduce chloride ion penetration. In those cases, and assuming that severe corrosion was experienced late in the life of the structure, the IR potential drop in the region below the waterline may conceivably be too large to be overcome by the modest working potential of a galvanic anode.

Figures 3a and 3b show (for x= 10 cm) the effect of variations in the oxygen diffusivity on the extent of protection achieved. Two steel positions (center of the top and the bottom of the rebar cage, or points D and A, respectively in Figure 2) were chosen for the comparisons. Each diagram shows the value of the unprotected corrosion current i_u for each diffusion coefficient, and the value of the corrosion current after protection as a function of the concrete resistivity. Good levels of protection were achieved both at the top and the bottom at moderate to low resistivities when D was the smallest (10^{-5} cm^2/sec). However, this benefit resulted because the unprotected corrosion currents were very small (less than 3 10^{-7} A/cm^2). Conversely, when D was high (10^{-3} cm^2/sec) the unprotected currents were high (several μA/cm^2) and significant protection could only be achieved at the bottom location (where resistive potential drops were lowest) and even then only for the lowest concrete resistivities. The top-bottom trends observed for x=10 cm were essentially replicated in the calculations for x=5 and x=15 cm, with differences as shown below.

Figures 4a and 4b illustrate (for point D, Figure 2, at the top of the cage) the effect of varying x from 5 cm to 15 cm on the extent of protection achieved. For D= 10^{-4} cm^2/sec and 10^{-5} cm^2/sec the protection was more effective for the thicker concrete cover. This is as expected since the thicker cover coupled with sometimes nearly diffusion-limited conditions caused smaller unprotected corrosion currents, facilitating the cathodic polarization. However, for D=10^{-3} cm^2/sec the unprotected corrosion currents were large for both high and low cover values. The consequently high polarizing current demand resulted in severe resistive potential drops and little protective efficiency independent of the concrete cover. For positions at the bottom of the cage (Point A, Figure 2) the values of the ratio i_u/i_p did not vary greatly from those for the case of Figure 3a.

Table 3 shows the calculated total anode current demand for each of the cases considered.

DISCUSSION

The calculations presented here are a sensitivity analysis of the effects expected when applying cathodic protection to a marine substructure member located in the tidal or near-tidal zone. Several approximations and estimates were made to provide input parameters for the computations. Some of the parameters (ρ, D, x) were chosen to span a plausible range bracketed by extreme conditions that may be ruled out as new experimental information becomes available. Others, such as the kinetic constants for the corrosion reactions (as well as the choice of only a limited number of possible reactions) represented a likely approximation. The outcome of the calculations should be viewed only as a guide to the relative variations in performance to be expected when critical concrete and dimensional variables span a range of possibilities.

The assumption that ρ and D were independent of radial position and elevation was also imposed. In actuality, significant drying could exist in the exterior of the upper portion of the footer, while incomplete water saturation may be present at short distances from the concrete surface even in the submerged portion [Nilsson, 1996]. The resulting elevational and radial gradients in ρ and D complicate the corrosion distribution, and should be considered in computations beyond the exploratory analysis presented here. Furthermore, cases in which the steel cage is not fully active merit additional consideration.

With the above limitations, the calculations showed that protection with a submerged galvanic anode of an entire footer of usual dimensions may be feasible if concrete resistivity in the area near the high tide mark is very low (a few kΩ-cm). Resistivity values in that range have been measured in some bridges in the Florida Keys with concrete that also exhibited high values of the effective chloride diffusion coefficient. However, the calculations suggest that protection above the waterline (and in extreme instances, possibly even below the waterline) is of limited effectiveness when the concrete resistivity is representative of that observed in wet, medium to high quality concretes (10-100 kΩ-cm). Values within this range have been often observed in the tidal region of bridges built in Florida with concretes with fly ash cement replacement which also tend to show low chloride diffusion coefficients [Sagüés et al., 1994b]. This observation should be taken into account when contemplating the use of submerged galvanic anodes in future applications, as the use of concretes with pozzolanic additions increases.

The calculated performance was least sensitive to the assumed value of the concrete cover x, and highly sensitive to the assumed value of the diffusion coefficient of oxygen in concrete. Unfortunately, the value of D is one of less certain parameters in actual systems. In the absence of additional information, (and recognizing that the concrete may not be completely saturated either above or below the water line) the value of 10^{-4} cm^2/sec appears to be a reasonable, order of magnitude estimate of the average value of D in a footer in near-tidal conditions. The calculations using that value may be used as an initial predictive guide, supplemented by the results from the extreme D limits as an indication of the sensitivity of the protection system to this parameter. Obviously the exploratory calculations presented here can not yield precise information concerning the corrosion rate and distribution if D varies spatially, which is likely. For example, the presence of a fully water saturated concrete layer near the surface below water (Gjorv et al., 1986) could lower oxygen transport and i_u in the submerged portion to values much smaller than those presented in Figure 2 even if the average value of D remains the same. These and other potentially important factors, such as the presence of oxygen consuming biodeposits on the concrete surface, should be considered in future analysis.

Only one working anode potential (-1V vs. SCE) was considered in the calculations. As shown in Table 3, calculated current demand in all cases is relatively modest and within the range achievable with normal commercial anode arrangements. Small variations in the working potential due to current demand (within the recommended operating potential-current curve for a given anode) are expected to have relatively minor effects and no attempt was made to include this as a variable in the calculations.

This paper applies primarily to the question of using a submerged galvanic anode as the main protection approach for the footer. However, submerged anodes are often used in combination with other above-surface systems. In other galvanic protection schemes the submerged anode is intended mainly as a means of avoiding unnecessary current draw from surface anodes placed above the waterline to protect columns or piling. The performance of these combined systems has been considered elsewhere [Kessler et al., 1996].

CONCLUSIONS

1. Cathodic protection of a marine substructure footer of typical dimensions by means of a submerged zinc anode appears to be feasible when the concrete resistivity is low (2 kΩ-cm) and the concrete approaches water saturation.

2. The calculations suggest that protection above the waterline (and in extreme instances, possibly even below the waterline) is of limited effectiveness for medium to high (10 - 100 kΩ-cm) wet concrete resistivities.

ACKNOWLEDGMENTS

The authors acknowledge the support of the Florida Department of Transportation and Engineering Computing Services of the University of South Florida. The findings and opinions expressed here are those of the authors and not necessarily those of the supporting organizations. One of the authors (F.P-M.) acknowledges the scholarship provided by the National Council for Science and Technology (CONACYT-México)

REFERENCES

Andrade, C; Alonso, MC (1994): Values of Corrosion Rate of Steel in Concrete to Predict Service Life of Concrete Structures. In: Application of Accelerated Corrosion Test to Service Life Prediction of Materials; ASTM STP 1194 (Eds: Cragnolino, G; Sridhar, N) American Society for Testing and Materials, Philadelphia, 282-295.

Funahashi, M; Bushman ,J.B. (1991): Corrosion vol 47: p. 376

Gjorv, O; Vennesland,O; El-Busaidy, A (1986): MP vol 25, 12: p 39

Kessler, R.J.; Powers, R.G.; Lasa I.R. (1996): MP vol 35, 12: p 11

Kranc, S.C.; Sagüés, A.A. (1994): Corrosion vol 50: p 50

Nilsson, L. (1996): Moisture in Marine Concrete Structures - studies in the BMB-project 1992-1996. In: Durability of Concrete in Saline Environment, CEMENTA AB, Danderyd, Sweden.

Sagüés, A.A.; Powers, R.G. (1996): Corrosion vol 52: p 508

Sagüés, A.A. et al (1994a): Factors Controlling Corrosion of Steel-Reinforced Concrete Substructure in Seawater. Report FL/DOT/RC/0537-3523, National Technical Information Service, Springfield, VA 22161.

Sagüés, A.A. et al (1994b): Corrosion of Epoxy Coated Rebar in Florida Bridges. Final Report to Florida D.O.T. WPI No. 0510603.

Tuutti, K (1982): Corrosion of Steel in Concrete (Stockholm, Sweden: Swedish Cement and Concrete Research Institute)

Table 1. Kinetic constants for the corrosion reactions.

i_{0a} = 1.875 10^{-2} μA E_{0a}= -780 mV SCE β_a = 60 (mV/decade)

i_{0c} = 6.25 10^{-4} μA E_{0c}= 160 mV SCE β_c = 160 (mV/decade)

C_0= 2.5 10^{-7} mol / cm^3 (Assumed O_2 solubility, in pore water)

Table 2. Corrosion current density (μA/cm^2), before application of cathodic protection, at the center of the top of the rebar cage.
(Point D, Figure 2)

Concrete Cover (cm)	Resistivity kΩ-cm	O_2 Diffusion Coefficient cm^2/sec		
		10^{-3}	10^{-4}	10^{-5}
5	2	14.42	2.25	0.23
	20	14.42	2.25	0.23
	100	14.42	2.25	0.23
10	2	9.22	1.15	0.11
	20	9.22	1.15	0.12
	100	9.22	1.14	0.12
15	2	6.71	0.80	0.08
	20	6.71	0.77	0.08
	100	6.71	0.77	0.08

(Except at the edges (see Figure 2), the corrosion current density did not vary significantly over the entire rebar cage assembly)

Table 3. Steel Potential (mV) SCE, before application of cathodic protection, at the center of the top of the rebar cage. (Point D, Figure 2)

Concrete	Resistivity	Diffusion Coefficient cm^2/sec		
Cover (cm)	kΩ-cm	10^{-3}	10^{-4}	10^{-5}
	2	-607	-655	-715
5	20	-607	-655	-714
	100	-607	-655	-714
	2	-618	-673	-735
10	20	-618	-673	-732
	100	-618	-673	-732
	2	-627	-682	-741
15	20	-627	-683	-743
	100	-627	-683	-743

Table 4. Current Delivered by the Anode (A).

Concrete	Resistivity	Diffusion Coefficient cm^2/sec		
Cover (cm)	kΩ-cm	10^{-3}	10^{-4}	10^{-5}
	2	2.62	0.43	0.05
5	20	0.43	0.27	0.04
	100	0.10	0.08	0.03
	2	1.32	0.23	0.03
10	20	0.20	0.13	0.02
	100	0.04	0.03	0.01
	2	0.87	0.15	0.02
15	20	0.12	0.08	0.01
	100	0.03	0.02	0.01

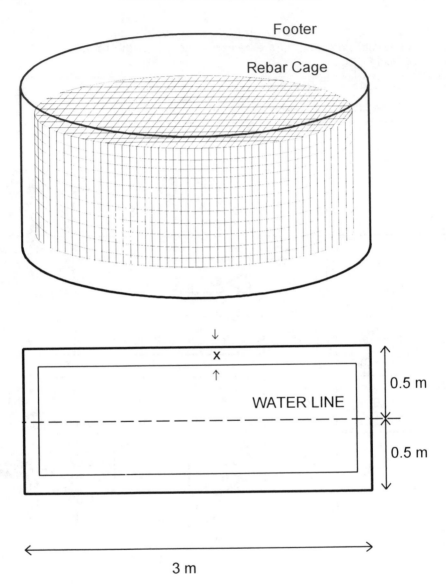

Figure 1. Geometric configuration of the footer system

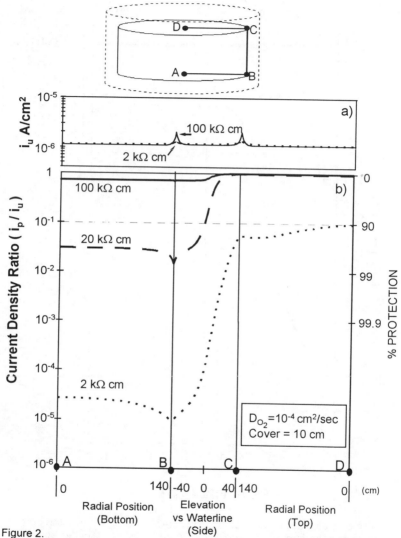

Figure 2.
  a) Calculated corrosion current density i_u as function of position in the rebar cage before application of cathodic protection for the cases with x=10 cm and D=10^{-4} cm^2/sec.
  b) Left ordinate: Ratio i_p/i_u (corrosion current densities after/before application of cathodic protection). Right ordinate axis: corresponding % protection; the 90% protection level is indicated by the dashed horizontal line.

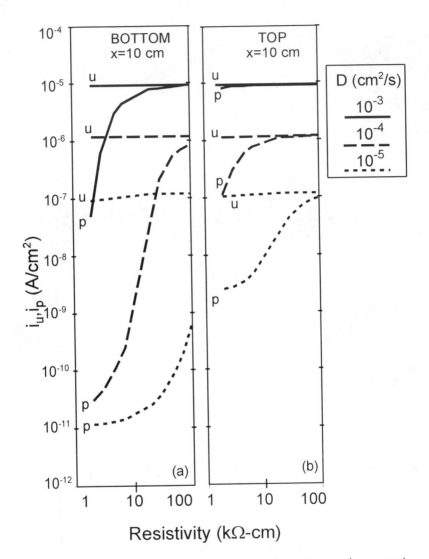

Figure 3. Calculated effect of cathodic protection on corrosion current density at a) bottom center (point A, Figure 2), and b) top center (point D, Figure 2) of the rebar cage as function of concrete resistivity and oxygen diffusion coefficient for a fixed concrete cover x=10 cm. The labels "u" and "p" indicate the corrosion current densities before and after application of cathodic protection, respectively.

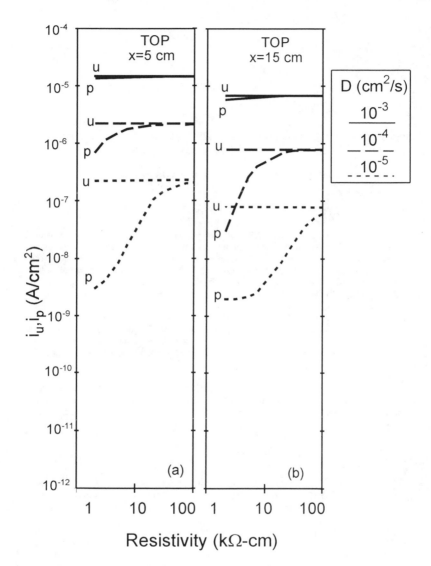

Figure 4. Calculated effect of variation of concrete cover x, illustrated for the at top center position of the rebar cage. See Figure 3 caption for legend designation.

Practical Application of Cathodic Protection Systems for ReinforcingSteel Substructures in Marine Environment

Ivan R. Lasa, Rodney G. Powers, Richard J. Kessler [1]

Abstract

With more than 1,903 km (1,200 miles) of coastline, the State of Florida in the United States has approximately 3,000 concrete bridge structures located in a corrosive marine environment. Under these conditions, the time for the steel reinforcement in the concrete to exhibit initial signs of corrosion is within two to three decades. In recent years, the Florida Department of Transportation (FDOT) has installed cathodic protection systems on sixteen bridges to attenuate concrete deterioration due to corrosion. These cathodic protection systems include impressed current as well as sacrificial anode systems. The impressed current group includes: 1) conductive mastic anode, 2) conductive rubber anode, 3) titanium mesh anode embedded in gunite and structural jackets, and 4) titanium mesh anode embedded in standard pile jackets. The sacrificial group includes: 1) arc-sprayed zinc anode, 2) expanded zinc sheet anode, 3) submerged bulk zinc anodes, and 4) expanded zinc mesh anode embedded in standard pile jackets.

This paper describes the cathodic protection technique and discusses the practical applications, installation, and performance of these cathodic protection systems as a means for controlling corrosion in steel reinforced concrete structures in the marine environment.

Introduction

Corrosion

Corrosion of reinforcing steel in concrete components is a problem of major concern when designing structures to be located in or near the marine environment. It is also of concern in structures that will be exposed to de-icing salts or other corrosive media.

Under normal conditions concrete is an alkaline material with a pH of about 12.5 due to the presence of calcium hydroxide[2]. At this pH value a passivating film forms around the reinforcement inhibiting any corrosion activity. As the pH of the concrete surrounding the reinforcement is reduced as a result of chloride intrusion or carbonation, the protecting film around the reinforcement is deteriorated and corrosion activity develops (Figure 1).

---

[1] *Florida Department of Transportation, Corrosion Research Laboratory, 2006 N.E. Waldo Road, Gainesville, Florida 32609*

The effects of reinforcement corrosion in concrete include cracking, and spalling. Corrosion of the reinforcing steel also contributes to the loss of cross-sectional area of the steel, and reduces the bond between the steel and the concrete[15]. The loss of cross-sectional area and the composite action between the steel and the concrete produces a loss of stiffness in the structural component and a decrease in the ultimate load capacity. If this cycle and the reaction are allowed to continue, the structure could eventually fail.

Conventional non-electrochemical rehabilitation techniques such as patching, gunite, or jacketing offer only a temporary remedy since these methods only address the corrosion consequences (concrete spalls and cracking) and not the corrosion problem itself. The only rehabilitation method proven to mitigate corrosion activity of steel reinforcement in concrete is cathodic protection[6].

Cathodic Protection

The cathodic protection concept is based on the capability of an applied current discharged by an anode to polarize the metal to be protected. The current flow forced from the externally placed anode polarizes the metal in the negative direction thereby reducing the rate of the oxidation reaction to a neglible value[5].

Cathodic protection controls the corrosion of steel in concrete by applying an external source of direct current to the reinforcing steel through the concrete (Figure 2). It provides an external energy to the steel surface to inhibit the development and progression of corrosion. Since corrosion is an electrochemical reaction by nature, by controlling the flow of current, corrosion can be controlled.

There are two types of cathodic protection systems, a) impressed current systems, and b) galvanic anode systems. The impressed current system utilizes an external power source to provide the current discharged by the external anode onto the cathodically protected metal. The galvanic system utilizes a metal higher in electro-potential in relation to the metal being protected to produce the protective current.

In the mid 70's, due to the high cost of periodic maintenance repairs, and in an effort to identify an effective repair method which would prevent further loss of original concrete on structures subject to corrosion deterioration, the Florida Department of Transportation began experimenting with cathodic protection. In recent years the FDOT has installed cathodic protection systems on sixteen bridges to attenuate concrete deterioration due to corrosion progression. These systems include impressed current as well as galvanic anode systems. Some of these systems have become standard FDOT structure rehabilitation methods.

Initial Energizing and Performance

When energizing a cathodic protection system, it is necessary to measure the natural (static) voltage potential of the metal to be protected prior to applying the protection current. This natural potential is used as a reference point to determine the amount of current necessary to achieve successful cathodic protection. When current is applied by the system, the voltage potential moves in the negative direction further polarizing the metal to a potential cathodic in relation to the externally placed anode. Typically, a polarization of 100 millivolts in the negative direction provides effective cathodic protection. Table 1 shows the general potential polarization and operational current for the systems discussed in this paper.

Performance of these systems is measured by monitoring the voltage potential of the reinforcing steel under cathodic protection. Criteria for determining the successful performance include the 100 millivolt polarization/decay criterion[10], and the E Log I criterion. Of these, the most common and more practical is the 100 millivolt polarization/decay test. However, for experimental systems, the E log I criterion is considered more adequate since it provides information necessary to analyze and understand the electrical behavior of the system.

The polarization decay test criterion assumes that when interrupting the current of a cathodic protection circuit, the voltage potential of the metal under protection will depolarize toward its natural voltage. At this point the voltage decay can be measured, and polarization confirmed by means of simple mathematical computation. Although the rate of decay may vary from structure to structure (Figure 3), a decay of 100 mV is typically considered a confirmation of successful cathodic protection. Depolarization test results for the systems discussed in this paper are found in previously published literature[6,7,8,11].

The E log I criterion is based on the accurate measurement of the polarized voltage potential of the metal to be cathodically protected as current is applied. In this test, current is applied to the metal in small increments at predetermined intervals while the potential response is measured for each increase in current. The resulting voltage potential values are plotted against the logarithm of the applied current for each value and a curve representing the electro-potential characteristics of the system is establish. From the curve, parameters such as corrosion current, and cathodic protection minimum potential and current can be determined (Figure 4).

Experimental Systems

Conductive Coating System

In 1984, the first cathodic protection system was evaluated by the FDOT. The system was installed on standard piles, beams, and underdeck of two bridges on the east coast of the State. The system consisted of a mastic paint anode impregnated with a large amount of carbon to enhance its conductivity. The anode mastic paint was applied over the concrete surface and connected to a rectifier to provide the cathodic protection current.

On the piles, the mastic was applied to a height of 1.3 m (4 ft) starting from the high tide elevation since the mastic could not be applied over wet concrete surfaces. On beams and underdeck, the entire concrete surface intended to be protected was coated. The rectifier was installed on a central location of the bridge and wires were routed in conduit to each cathodic protection zone.

Since no established criteria to evaluate cathodic protection performance on reinforcing steel was available at that time, existing criteria used for pipelines were adopted. Upon energizing, an initial change in the voltage potential of the reinforcing steel was observed. The initial steel potential shifts obtained met the pre-determined cathodic protection criterion of -0.850 V (Cu/CuSO$_4$). On all beam and underdeck zones, cathodic protection was achieved. However, on the piles and components in direct contact with the tidal waters, the system was unable to constantly maintain the potentials at an acceptable cathodic protection level fluctuating below the -0.850 V. It was observed that the tidal changes affected the system's current distribution as the current was redirected to the submerged portion of the piles when the high tide waters came in contact with the anode. The mastic on these areas promptly disbonded from the concrete producing a non-uniform current distribution along the intended protected areas. The system was maintained in service and closely monitored for a period of seven years after which the structure was scheduled for replacement due to functional deficiencies.

Conductive Rubber Anode

Based on the previously discussed experience, an anode capable of uniform current distribution and which could be installed in direct contact with the water was developed for bridge pilings. The anode was developed consisting of a mat of rubber loaded with black carbon to produce a total volume resistivity of 1.5 ohm-cm. The rubber anode is provided on one side with grooves to allow flushing of excessive salt and debris which could accumulate at the anode-concrete interface. At the same time, the grooves allow the accumulation of moisture to enhance the concrete-anode interfacial electrical conductance.

The anode is 1.2 m (4 ft) long and is placed on the piles centered at mean high tide elevation (Figure 5). Length and width can be easily modified based on requirements dictated by piling dimensions and tidal changes although a length of 1.2 m (4 ft) is apparently sufficient in most cases. The anode is mechanically attached to all faces of the pile with the grooved side against the concrete surface using four fiberglass or recycled plastic compression panels with a silicon based rubber pad placed between the conductive rubber and the panels. The soft silicon rubber pad allows the conductive rubber to conform to concrete surface irregularities. The components are attached together with adhesive, and secured in place on the pile with five 1.9 cm (3/4 in). wide stainless steel bands. A rectifier is placed at a convenient location on the bridge and wires are routed in conduit to each cathodic protection zone.

The first system of this type was installed for evaluation in 1987 on ten piles on two bents (Bents 4N and 4S) at the B. B. McCormick Bridge in Jacksonville, FL with funding provided by the U.S. Federal Highway Administration (FHWA) Demonstration Projects Division. A constant voltage rectifier was initially used to produce the system current but for evaluation purposes, it was later replaced by a constant current-limiting voltage unit. The new rectifier produced the predetermined constant current during initial polarization while reducing it as the current demand was decreased after polarization was achieved by means of the preset limiting voltage. The system was energized using the E Log I criterion which determines the necessary amount of current to provide effective cathodic protection.

The E Log I test results indicated a required protection current of 0.730 A with a corresponding potential of -0.710 V (Cu/CuSO$_4$) for Bent 4N. Bent 4S was energized at a later time using E log I values of 0.284 A with a corresponding potential of -0.359 V. Bent 4N was energized at the above values using the constant voltage rectifier. Upon replacing the rectifier with the constant current-limiting voltage rectifier, the current was set at 0.530 A maintaining the polarized cathodic protection potential of -0.710 V (Cu/CuSO$_4$). Bent 4S was initially energized using the constant current-limiting voltage rectifier. Periodic monitoring over a three year evaluation period indicated an average voltage potential polarization of 384 mV from the static value. It was observed that after cathodic protection polarization occurred in the splash area, gradual polarization of the submerged portion of the pile was achieved [8] as indicated by potential measurements obtained with the reference electrode placed at elevations below the anode. In addition to the field evaluation, laboratory tests were conducted to determine the average efficient life of the system. Initial results indicated a service life between five and seventeen years[12]. On-going long term laboratory testing suggests a service life in excess of twenty years.

This system is commercially available and similar systems have been installed at the Ribault River Bridge in Jacksonville, Fl. and the Howard Frankland Bridge in Tampa, FL. This system is recommended for pilings exhibiting initial signs of corrosion deterioration (voltage potentials more negative than -0.350 V Cu/CuSO$_4$ and/or initial cracking). Since the anode requires a uniform concrete surface, adequate concrete restoration is required on piles with spalled concrete.

Titanium Mesh Anode Embedded in Gunite

The titanium anode mesh is an expanded catalyzed titanium mesh with an external coating of mixed metal oxides capable of current outputs of up to 33.4 mA/m^2 (3.1 mA/ft^2) of protected concrete without adversely affecting its estimated service life. The anode mesh is installed on the structure by attaching the mesh directly to the existing concrete surface at the areas of the structure to be protected using plastic fasteners. The mesh is provided in 1.2 m (4 ft) wide rolls which can be spliced together to larger widths by resistance welding two mesh panels over a strip of titanium bar. A strip of the titanium bar extends outside the protected area for connection to the wires originating at the rectifier. Following installation on the concrete, the anode is embedded in gunite at a depth of 5.1 cm (2 in).

The first system of this type evaluated by FDOT was installed in 1988 on a bridge pier at the Howard Frankland Bridge in Tampa, FL [7]. The pier is comprised of three square footer type pile caps, three rectangular columns, and two struts (Figure 6). The bottom portion of the footers come in direct contact with the tidal waters during the periods of high tide while columns and struts typically remain in a mostly dry condition. Although the system was designed as a single circuit system, reference electrodes were provided for each pier component (footers, columns, and struts) such that potentials could be monitored individually for each zone. This system was designed by the anode manufacturer who additionally provided quality control during construction.

After installation, the system was energized using the E Log I criterion. Static potentials for Zone 1 (columns), Zone 2 (struts), and Zone 3 (footers) were measured at -0.290, -0.441, and -0.464 V respectively. All potentials were measured vs the Ag/AgCl electrodes embedded in the concrete for each zone. Upon energizing, the polarized potentials were measured at -0.357, -0.490, and -0.570 V (Ag/AgCl) respectively. Two weeks after initial energizing, the polarized potentials measured -0.460, -0.629, and -0.680 V (Ag/AgCl) with a resulting polarization from static ranging from 170 mV in Zone 1 to 216 mV in Zone 3.

Voltage potential data gathered during the entire evaluation period indicated satisfactory cathodic protection levels even though after the first six months, the gunite placed at the elevation in direct contact with the water, partially delaminated from the original concrete surface. The delamination was attributed to the physical properties of the gunite material which could not provide a good bond to the existing concrete surface in moisture saturated conditions. It was determined that even though partial delamination of the gunite occurred, protection current was being discharged through the salt water since the reinforcement potentials were not affected.

Additional testing of the cathodic protection system included 4 hours depolarization tests which produced polarization decays ranging from 132 to 214 mV. At this time, this type of system is recommended for bridge components (typically mass concrete components) not in direct contact with tidal waters.

Titanium Mesh Encapsulated in Structural Jacket

The anode used for this system is a catalyzed expanded titanium mesh anode identical to that described in the previous section. This system is typically used on bridge mass concrete components which require structural rehabilitation due to degradation related to inadequate reinforcement, settlement, and/or corrosion. The system combines structural rehabilitation with corrosion control. Installation of the system requires the removal of all existing delaminated concrete and cleaning of the remaining concrete surface and exposed reinforcing steel. If present, remaining cracks in sound concrete are filled with mortar grout and the catalyzed titanium mesh anode is installed over the concrete surface. Adequate reinforcing steel encasement is then placed around the structure components to be encapsulated as dictated by the required structural repair. Forms for the new structural jacket are placed and the concrete is cast around the deteriorated component. Electrical isolation between the existing and the new reinforcing steel is always maintained such that at energizing, the two steel systems may be energized at different current levels since the older steel will exhibit a higher level of corrosion activity due to it's surrounding concrete.

The first of these systems was installed at Verle Allen Pope Bridge in Crescent Beach, Florida. The system was installed on eight pier footers (Figure 7) which had experienced structural deterioration as a result of inadequate reinforcement volume. The resulting cracks allowed for the intrusion of salt water into the concrete which consequently produced severe reinforcement corrosion and subsequent concrete spalling. The system design provided one constant current-constant voltage rectifier per pier (two footers). Connection wires to the steel were provided such that new and existing steel could be energized separately. Only the original existing steel was energized initially, and the corrosion level on the new steel was periodically monitored for connection to the system when needed.

When installation was completed, the system was energized by FDOT personnel using the E Log I criterion and performance was closely monitored. Static voltage potentials for the four systems ranged from -0.523 to -0.603 V (Cu/CuSO$_4$) and the E log I test results determined cathodic protection potentials shifts of around 100 mV on all systems. Two weeks after energizing, the voltage potentials had polarized to values ranging from 0.273 to 0.428 V from the static value. After two years of evaluation, the system was found to be very effective in forcing the current flow from the anode to the reinforcing steel surfaces as demonstrated by an average long term polarized potential of 900 mV [3]. Interestingly, the structural cracks product of the inadequate reinforcement volume re-appeared on one of the footers soon after the repairs were completed allowing salt water intrusion to the reinforcement. For three years, at which time the footer was replaced due to structural concerns, the cathodic protection system maintained the reinforcing steel at a corrosion free level, which was verified by visual inspection when the footer was demolished for replacement and the existing rebars extracted.

Following the success on this bridge, a similar system was incorporated to repairs on two more bridges. On these bridges, the system was combined with structural repairs requiring structural jackets with post-tensioning. In all cases, the system has proven very effective in controlling corrosion.

Impressed Current CP Pile Jacket System

This system is specifically designed for corrosion control on bridge piles. Like the previously discussed systems, it is of the impressed current type and requires an external power supply to provide the cathodic protection current. The anode utilized by the system is the expanded titanium mesh anode which is mechanically suspended on the inside face of a standard fiberglass stay-in-place form (Figure 8). The fiberglass form is placed around the pile at low tide elevation and typically extends 1.8 m (6 ft) upward. The length of the jacket can be specified based on the existing deterioration. The jacket provides a uniform annular space between the form and the existing concrete surface of 7.6 cm (3 in) which is filled with portland cement-sand mortar. Any deterioration outside the CP jacket limits is repaired using a good quality mortar/concrete.

The concept was developed by the Florida Department of Transportation and the first system was installed on forty-four piles at the Ribault River Bridge in Jacksonville, FL. On this project the jackets were 1.2 m (4 ft) tall and were placed centered at high tide elevation. The system included four rectifiers installed at different locations on the bridge with the capability to provide individual current output adjustment to each pile bent. Installation of the jackets consisted of removing all delaminated concrete and cleaning the remaining concrete surface and exposed steel of all marine growth or debris. Because the jackets were later filled with mortar, no concrete restoration was required. The anode jackets were installed and filled as specified, and a conduit-wiring system was installed to provide the steel and anode connections to the rectifiers.

The system was energized using the E Log I criterion on four bents, and the -850 mV criterion on the remaining [9]. The initial current density at the anode ranged from 9 to 22 mA/m^2 (0.8 to 2.0 mA/ft^2) with a steady-state current after polarizing of 6.4 mA/m^2 (0.6 mA/ft^2). Polarized voltage potentials ranged from -0.780 to -0.990 V (Cu/CuSO$_4$). In both cases, effective cathodic protection was achieved as demonstrated by periodic potential depolarization tests, which suggested a minimum polarization of 150 mV on each bent. This system was later installed on two other bridges and likewise exhibit excellent performance. This system has become one of the FDOT standard bridge pile rehabilitation methods.

Sprayed Sacrificial Zinc Metalizing System

This system is of the sacrificial anode group. It uses the zinc anode which is higher in electro-potential in relation to the steel, to provide the cathodic protection current. The potential of the zinc anode is around -1.1 volts while the potential of corroding steel is around -0.450 volts (Cu/CuSO$_4$). The installation consists of removing all the deteriorated concrete from the structure and sandblasting clean the remaining concrete surface and exposed steel. The zinc is applied over the concrete surface as well as the exposed reinforcement. Spraying over the reinforcement provides the electrical connection between the zinc and the steel. In this manner, the zinc directly protects the exposed steel while the steel within the concrete receives cathodic protection current through the concrete itself (Figure 9).

Application of the zinc anode is made similarly to spray painting. A hand held gun creates an electric arc between two zinc wires fed through the gun. The zinc melts at this point and is sprayed onto the concrete surface by an air nozzle provided at the gun. The zinc coating is applied to a thickness ranging from 0.38 to 0.5 mm (15 to 20 mils). The typical zinc to concrete bond strength is around 1,034 kPa (150 psi). This system has the capability of functioning as a sacrificial or impressed current system (no direct contact to the steel if used as impressed current), although the FDOT has only used it in sacrificial mode.

Initial evaluation of this system was conducted on five circular columns at the Niles Channel Bridge in the FL. Keys in 1989. The columns were 0.9 m (3 ft) in diameter and were built using epoxy-coated rebar. All of the columns exhibited advanced stages of corrosion development with significant spalling. Three of the columns were metalized and further visually evaluated and sound tested periodically to observe any corrosion progression. The other two columns were instrumented prior to metalizing such that current flow and polarization could be measured. Polarization measurements in excess of 100 mV were observed at all elevations within the metalized area on both instrumented columns. The protective current delivery of the system has been maintained at about 0.5 uA/cm^2 (0.5 mA/ft^2) [13]. After five years, only one bar on one of the columns exhibited some degree of corrosion progression.

The estimated service life of the anode is between seven to eight years at which time re-metalizing will be required. This system is recommended for applications not in direct contact with the water since this accelerates the consumption rate of the anode, and significantly decreases the anode service life. Additional evaluation has been performed on structures containing standard (uncoated) rebar. Protective current delivery on standard bars was around 1.1 uA/cm^2 (1 mA/ft^2) and polarization in excess of 100 mV was produced in all cases[14]. At this time FDOT has used this system in over eight bridges comprising over 18,400 m^2 (200,000 ft^2) of metalized concrete. The overall successful performance of this system is around 95 percent.

Perforated Zinc Sheet System

This sacrificial cathodic protection system was designed to provide corrosion protection to bridge concrete pilings at splash zone elevations. The system consists of perforated zinc sheets which are firmly held against the concrete surface and mechanically connected to the reinforcing steel. The zinc sheets are commercially available and conforms to ASTM A-190 with a chemical composition essentially 99.9 percent pure zinc. The sheets have a weight of 7.9 kg/m^2 (1.58 lb/ft^2) with a density of 0.02 kg/in^2 (0.259 lb/in^2). The sheets are fabricated into an open cage of sufficient size to be wrapped around the four faces of the concrete pile. The cage is then secured in place by means of four specially designed, recycled wood/plastic panels compressed firmly against the pile with five 316 grade stainless steel bands. The panels are made of 50% plastic and 50% wood fiber and are provided with grooves on the inside face to allow for moisture accumulation and to washout resulting zinc oxides. The zinc sheet anode is then connected to the reinforcing steel via a copper strand wire or other suitable connection.

The system is installed on the piles centered at high tide elevation (Figure 10). In order to provide protection to the submerged portion of the pile, a 21.8 kg (48 lb) bulk zinc anode is attached to the pile at an elevation of 0.6 m (2 ft) below low tide. The primary function of this secondary anode is to polarize the submerged portion of the pile, therefore preventing this area of the reinforcing steel from attracting current from the perforated zinc anode during high tide periods when the water contacts the perforated anode. The bulk anode is connected to the pile at the same location as the zinc sheet anode via a No. 6 copper wire.

The first field evaluation of this system was conducted on ten piles at the B.B. McCormick Bridge in Jacksonville, FL. All of the piles were instrumented to allow measurement of the current and the voltage potential of the reinforcing steel. Two piles were instrumented to measure anode and steel current densities at various elevations at 0.305 m (1 ft) increments. Protective current densities measured ranged from 8.6 to 17.0 mA/m^2 (0.8 to 1.58 mA/ft^2) with the highest current densities at the lower elevations. Voltage potentials on the piles were polarized in the negative direction to levels ranging from 300 to 430 mV from static measured at elevations around the high tide level [4]. The evaluation concluded that the magnitude of the current produced by the zinc anodes overpowers the corrosion currents discharging from the anodic areas as demonstrated by the measured polarization, therefore providing effective cathodic protection. Additionally, by using recycled materials, this system is low cost while at the same time requires minimal maintenance thereby offering an attractive alternative to more complex impressed current systems.

Sacrificial Cathodic Protection Pile Jacket

This sacrificial anode cathodic protection system was developed by the Florida Department of Transportation to provide corrosion protection to bridge pilings where concrete restoration due to corrosion deterioration is required. The system protects the submerged portion of the piling, the splash area, and the area immediately above. The system consists of a standard pile jacket provided with an internally placed expanded zinc mesh anode, an additional bulk zinc anode installed at an elevation of 0.6 m (2 ft) below low tide, and a connection to the reinforcement (Figure 11). If necessary, sprayed zinc can be applied to the areas above the jacket to control any corrosion activity at this elevation. The jacket is a two piece, stay-in place fiberglass form provided with the expanded zinc mesh anode pre-installed on the inside face of the form.

Installation consists of removing all delaminated concrete from the pile and clean-up of the remaining concrete as well as the exposed steel. The jacket is then placed around the pile from low water elevation upward providing a 5.1 cm (2 in) annular space between the pile surface and the fiberglass form. The bulk anode is installed below water level at the specified elevation and the connection cable is routed upward inside the jacket to the anode steel connection location. The jacket is then mortar filled and the zinc mesh and bulk zinc anodes are connected to the reinforcing steel. The filling material is a portland cement-sand grout with a minimum cement content of 558 kg/m^3 (940 lb/yd^3) of grout.

Initial field evaluation was conducted by providing two standard reinforcement concrete piles at the Broward River Bridge in Jacksonville, FL with the system. The piles were provided with instrumentation to measure system current and the voltage potential of the reinforcing steel. The NACE 100 mV polarization and polarization decay criteria were used to evaluate the cathodic protection performance. Upon energizing, the voltage potential of the reinforced steel shifted from an average static of -0.305 V to an average energized (on) potential of -0.408 V with a polarized value of -0.395 V. After 400 days, the potential increased to an average of -0.676 V with a polarized value of -0.533 V. A polarization decay test conducted on each pile at this time measured potential depolarization levels of 118 mV for pile A and 165 mV for pile B [6].

The cost of this system favorably compares with the cost of standard pile jackets. When compared to impressed current systems, this system has the advantage of requiring only minimal monitoring and maintenance since external power supplies are not required. It is also expected that the minimum effective service life of the system will be 45 years based on the consumption rate of the anodes. This system is commercially available for various pile sizes.

Summary

For over twenty years the Florida Department of Transportation has evaluated and utilized the cathodic protection technique to mitigate corrosion of reinforcing steel in concrete. At this time, the use of cathodic protection has evolved into a practical, internationally recognized method of controlling corrosion.

Unlike patching or jacketing, cathodic protection prevents further original concrete loss on structures due to corrosion progression.

Although no single cathodic protection system is adequate for all applications, the availability of several systems assures that a cathodic protection method is available for individual simple or complex applications.

References

1.   Barnhart R.A. "Official Policy Statement, " Federal Highway Administration, April, 1982.

2.   Clear K.C. "Chloride at the Threshold - Comments and Data" American Concrete Institute Convention, California, 1983.

3.   Kessler R.J., Powers R.G., Lasa I.R. "Titanium Mesh Encased in Structural Concrete," Florida Department of Transportation, Report No. 89-9A.

4.   Kessler R.J., Powers R.G., Lasa I.R. "Cathodic Protection Using Scrap and Recycled Materials," NACE Paper 555, 1991.

5.   Kessler R.J., Powers R.G. "Update in Cathodic Protection of Reinforcing Steel in Concrete Marine Structures," NACE Paper 326, 1992.

6.   Kessler R.J., Powers R.G., Lasa I.R. "Zinc Mesh Anodes Cast Into Concrete Pile Jackets," NACE Paper No. 327, 1996.

7.   Langley R., Powers R., Kessler R. "Impressed Current Cathodic Protection System on Howard Frankland Bridge'" Florida Department of Transportation, Report No 89-10A, 1989.

8.   Lasa I.R., Powers R.G., Kessler R.J., "Evaluation of Conductive Rubber, Perforated Zinc Sheet, and Titanium Mesh Cathodic Protection Systems for Bridge Pilings," Report for Federal Highway Administration, Contract No. DTFH71-86-923-FL-07.

9.   Martin B.L., Firlotte C.A. "Protecting Substructures in Marine Environments," NACE, Materials Performance Journal, volume 34, 1995.

10.   NACE Standard Recommended Practice RP0290-90 "Cathodic Protection of Reinforcing Steel in Atmospherically Exposed Concrete." 1990

11.   Powers R., Lasa I. "Impressed Current Cathodic Protection Using Titanium Mesh Cast in Structural Concrete," Florida Department of Transportation, Report No. 93-3A, 1993.

12.   Sagues A. "Laboratory Evaluation of Conductive Rubber Anode," University of South Florida, 1992.

13.   Sagues A., Powers R. "Low Cost Sprayed Zinc Galvanic Anode for Control of Corrosion of Reinforcing Steel in Marine Bridge substructures," Strategic Highway Research Program, Contract No. SHRP-88-ID024, 1993.

14.   Sagues A., Powers R. P. "Sprayed Zinc Galvanic Anode for Corrosion
       Protection of Marine Substructure Reinforced Concrete," Transportation
       Research Board National Research Council, Report NCHRP-92-ID003, 1995.

15.   Shalon R., Raphael M. "Influence of Seawater in Corrosion of   Reinforcement"
       ACI journal, vol. 55, 1959.

TABLES

Table 1:   General performance of cathodic protection systems

System	Service Period (Months)	Initial Current mA/m^2	Steady-State Current mA/m^2	Initial Polarization From Static mV	Long Term Polarization From Static mV
Conductive Rubber	60	58.1	10.7	130	425
Titanium Mesh in Gunite	48	21.5	13.9	81	380
Titanium Mesh in Structural Jacket	36	19.4	9.6	100	337
Titanium Mesh Pile Jacket	48	15.5	6.0	150	305
Sacrificial Sprayed Zinc	23	33.4	10.7	100	350
Perforated Zinc Sheet	12	29.9	12.9	114	385
Zinc Mesh Pile Jacket	34	15.1	9.7	90	228

* Values are average of several CP circuits per system (as applicable).

FIGURES

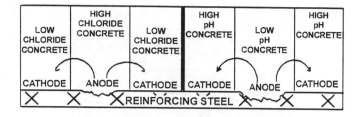

Figure 1:   *General corrosion mechanism of reinforcing steel in concrete.*

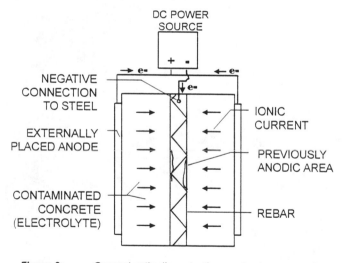

Figure 2:   *General cathodic protection mechanism.*

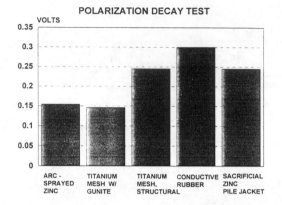

Figure 3: Measured polarization decay of the reinforcing steel for various cathodic protection systems.

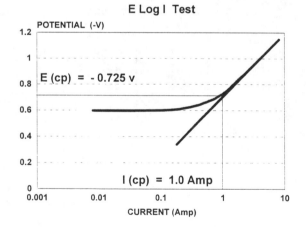

Figure 4: Typical E log I test curve.

Figure 5: First Conductive Rubber anode system at B.B. McCormick Bridge.

Figure 6: Titanium mesh anode embedded in gunite at Howard Frankland Bridge.

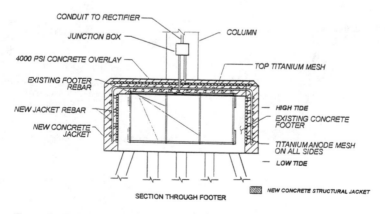

CONDUIT TO RECTIFIER

JUNCTION BOX

COLUMN

4000 PSI CONCRETE OVERLAY

TOP TITANIUM MESH

EXISTING FOOTER REBAR

NEW JACKET REBAR

HIGH TIDE

EXISTING CONCRETE FOOTER

NEW CONCRETE JACKET

TITANIUM ANODE MESH ON ALL SIDES

LOW TIDE

NEW CONCRETE STRUCTURAL JACKET

SECTION THROUGH FOOTER

Figure 7: Titanium mesh anode embedded in a reinforced concrete structural jacket.

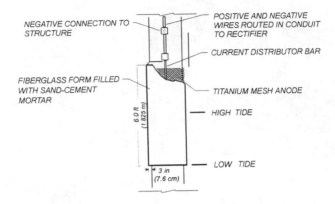

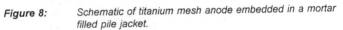

Figure 8:   Schematic of titanium mesh anode embedded in a mortar filled pile jacket.

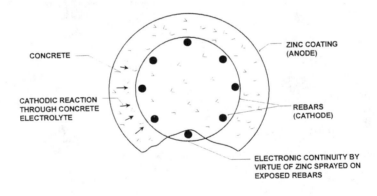

Figure 9: Schematic of sacrificial arc-sprayed zinc cathodic protection on a circular column.

Figure 10A: Perforated zinc sheet anode prior to panel installation.

Figure 10B: Perforated zinc sheet anode with recycled wood-plastic compresion pannels

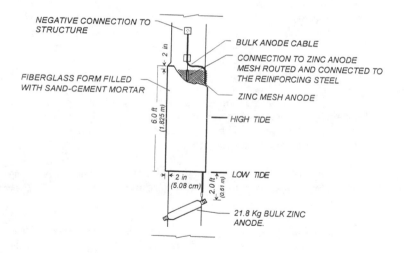

NEGATIVE CONNECTION TO STRUCTURE

2 in

BULK ANODE CABLE

CONNECTION TO ZINC ANODE MESH ROUTED AND CONNECTED TO THE REINFORCING STEEL

FIBERGLASS FORM FILLED WITH SAND-CEMENT MORTAR

ZINC MESH ANODE

6.0 ft (1.825 m)

HIGH TIDE

2 in (5.08 cm)

2.0 ft (0.61 m)

LOW TIDE

21.8 Kg BULK ZINC ANODE.

Figure 11: Schematic of sacrificial cathodic protection pile jacket with expanded zinc mesh.

Calculation of the Residual Life of Postensioned Tendons Affected by Chloride Corrosion Based in Corrosion Rate Measurements

by C. Andrade[1], C. Alonso[2], J. Sarría[3] and A. Arteaga[4]

Abstract

Residual Life of corroding concrete structures is being studied because of the economical consequences of the need of repairing. The on-site measurement of the corrosion rate, CR, is the most useful tool up to the moment for predicting the evolution of the damage and its structural consequences.

In the present paper the methodology of calculation of residual life from corrosion rate values is illustrated by means of an example of application to the case of a postensioned bridge. Chlorides were added by mistake to the mixing water of this bridge which started to corrode from the beginning of its life. Since this fact was carried out a periodical measurement of the corrosion current I_{corr} in critical points of the bridge was carried out.

The corrosion monitoring was made by using a corrosion-rate-meter having a guard ring. The method used for the, CR, measurements was based in the Polarization Resistance.

The results indicated that the CR was relatively small, varying from point to point and along the time. Also was noticed that the deck was much drier in the center than in the lateral parts. An statistical treatment of the results has been made

---

[1] Dr. Industrial Chemistry, Institute "Eduardo Torroja", CSIC, Serrano Galvache , s/n. 28033. Madrid. Spain.

[2] Dr. Chemistry, Institute "Eduardo Torroja", CSIC, Serrano Galvache , s/n. 28033. Madrid. Spain.

[3] Architect, Institute "Eduardo Torroja", CSIC, Serrano Galvache , s/n. 28033. Madrid. Spain.

[4] Dr. Civil Eng. , Institute "Eduardo Torroja", CSIC, Serrano Galvache , s/n. 28033. Madrid. Spain.

in order to establish a representative corrosion rate value, I_{REP}, which enables the prediction of the time to failure of the tendons.

Introduction

Prediction of residual life is focusing increasing interest due to the progressive deterioration of the concrete structures. The main cause of this deterioration is the penetration of chlorides into concrete (Hausmann, 1967), due to the structures are placed in marine environments or when deicing salts are used on them.

The corrosion, developed when enough amount of chlorides reach the reinforcement, leads to several damages: cover cracking, loss in bond and in bar cross-section. The residual life calculation tries to quantify the rate of deterioration of the element or of the whole structure.

Due to the aggressive character of the chlorides, national codes prescribe the avoidance of these ions in the mixing water in quantities beyond a certain limit or threshold. This limit is made lower in prestressed than in reinforced concrete. In Spain the requirement for prestressing wires specifies a limit of 250 ppm of chloride ion in the mixing water [EP-93].

However, in spite of this prescription being usually strictly respected an accident happened. It was, due to an unexpected sudden salinization of a well, whose water was used during the erection of a postensioned bridge. Thus, the deck and grouting paste were contaminated with chlorides in quantities ranging between 1000-2000 ppm.

The salinization occurred due to the long dry climatic period suffered during the last years by large regions of Spain. This dry period without rains induced the presence of increasing chloride contents in the tap water, mainly in coastal areas, where sea water permeated into the underground and contaminated the water table.

The particular condition of the bridge was immediately studied and corrosion rate measurements over more than two years were carried out. The detailed study has been aimed to calculate the residual life of the bridge. In the present paper the work carried out is presented in order to illustrate a methodology of calculation of residual life based in the measurement of the corrosion rate along time and its further statistical treatment.

Bridge description and damages

The bridge was erected between summer 1991 and June 1992. It has a mixed section with a metallic girder, the deck being postensioned. The total length is of

316 m composed of five spaces (four piers) of 40, 68, 100, 68 and 40 m. The deck is about 30 m wide and 25 cm thick, with a constant section all along. A section of the bridge is shown in figure 1.

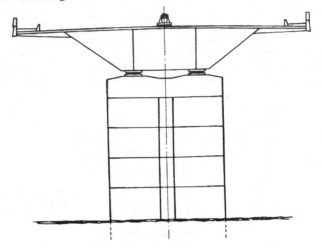

Fig. 1 A section of the deck and girder and a pier of the bridge.

The presence of chloride in the mixing water was noticed when concreting the last span. Immediately a visual inspection was undertaken. Several perforations to reach the reinforcement and tendons were carried out and some spots of rust were detected confirming the initiation of a localized corrosion. Any other apparent damage appeared.

The decisions undertaken were then: 1) to reinforce the deck by means of an additional top heavy reinforced concrete layer, 2) the introduction of additional nonadherent cables running along the interior of the girder. These cables are prepared to be stressed in three phases: one immediately and the other two, in function of the gradual section loss induced by the corrosion and 3) to submit the bridge to continuos corrosion monitoring, in order to help the decision of progressive prestressing.

Inspection methodology of corrosion

Testing zones

The deck was first divided in testing zones. These zones were selected taking into account: 1) The concreting phases, 2) The concentration of stresses, 3) The presence of joints.

Thirty testing points were finally selected, 12 of them in joints between spans. In these zones, perforations were made in order to reach the duct for direct visual inspection. The holes are kept permanently open in order to facilitate the electrical contact with the reinforcements.

Measurements

An initial visual inspection was made on the normal reinforcements, on the ducts and in the tendons (once the duct was broken). This visual observation was made at an age between 1-6 months after concreting.

In order to carry out a contrast of the electrochemical measurements, when the bridge was about one year old, in one of the inspection points, a measurement of the cross section loss, was also undertaken. In consequence, two wires of one tendon were cut, and the pit depths were measured by optical microscopy.

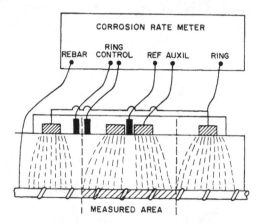

Fig. 2 Confinement of the current lines due to the use of a sensorized guard-ring.

The third type of measurement was the periodic monitoring of the corrosion rate. It was made by means of the equipment Gecor 06 (Feliú, 1990). This equipment is able to make a sensorized confinement of the electrical current applied, which is confirmed by a guard ring (see Figure 2). The method is based in the recording of the Polarization Resistance, R_p, which enables the calculation of the corrosion current, I_{corr} through Stern's formula (Stern, 1957), $I_{corr} = B/R_p$. A value of 26 mV is used for the constant B.

The electrical contact was made indistinctly to the normal reinforcements or to the duct, as it was determined that all the bars, ducts and tendons were electrically connected.

Assuming a generalized corrosion, the attack penetration can be calculated through Faraday's Law, by means of the following equivalence (Rodriguez, 1996):

$$x = 0.0116 \, I_{corr} \, t \qquad\qquad [1]$$

were x = attack penetration in mm, t = the time from steel depassivation and I_{corr} is the averaged value of corrosion intensity in $\mu A/cm^2$.

However, in present case the attack is localized and therefore the residual bar diameter Ø can be estimated (Rodriguez, 1996) as Figure 3 shows, from the nominal diameter $Ø_0$ by:

$$Ø = Ø_0 - \alpha x \qquad\qquad [2]$$

where α = pit concentration factor=10 in present case. The value of 10 was selected after contrast with the microscopic observation of the pit depth, as will be described later on.

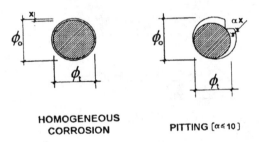

HOMOGENEOUS
CORROSION   PITTING [α≤10]

Fig. 3 Residual reinforcing bar section for homogeneous (left) and localized (right) attack.

Results

Visual inspection and determination of maximum pit depth rate

As was mentioned, direct observation of normal reinforcements, ducts and tendons inside the ducts, was carried out. It revealed that localized attack was developed in the form of small pits. The corrosion was similar in all metallic elements. The amount of chlorides in the concrete mass was of 0.083 % by mass of concrete and of 0.05 % by mass of grout.

Concerning the pit depth, it was measured from the cross section, of one of the wires that were cut for microscopic observation, (see figure 4) when the life of the structure was about 1 year old. The pit depths found range between 30 to

130μm, which is equivalent to a corrosion intensity of 30-130/11.5 = 2.7 to 11.8 μA/cm^2 . These values are around ten times those of the CR measured, which confirmed the α value of 10 introduced in equation [2].

Fig. 4 Magnified cross section of a tendon showing incipient corrosion

Corrosion rate measurements

A total of 30 points were prepared for measuring the CR. Figure 5 shows the evolution of Icorr along time in several of the measuring points.

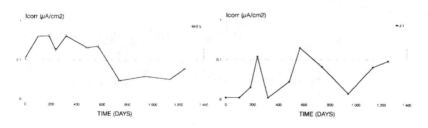

Fig. 5 Evolution of I_{corr} along time.

If an average of the I_{corr} values is made for each section or joint, the results obtained can be seen in Figure 6. This figure enables to deduce that the testing points of the section S5 to S10 and those in the joint J9 and J10, presents the highest I_{corr} values.

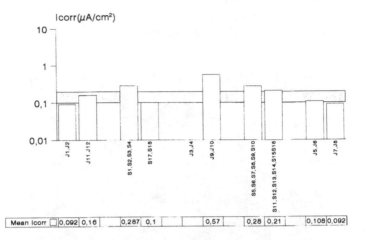

Fig. 6 Average Icorr in the different testing points.

Converting these I_{corr} values into attack penetration, V_{corr} (in μm/year) values around to 20-30 μm/year are obtained, that multiplied by a factor of 10, give a maximum pit depth of 200-300 μm/year, which largely correspond to the maximum pit depths measured by microscopic observation. This means that the local penetration of attack can be calculated by multiplying the average V_{corr} values in μm/year by a pit concentration factor of 10 times.

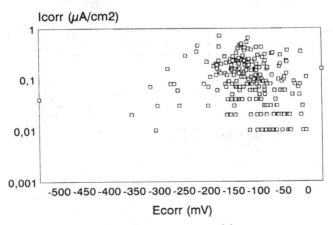

Fig. 7 Corrosion rate vs. potential.

In figure 7 is presented the relation found between the Icorr and Ecorr values. The lack of accordance is evident. Figure 8 shows the variations along time of the resistivity, relative humidity in the ambient of the girder and temperature.

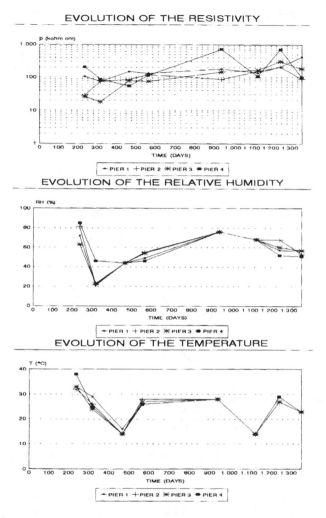

Fig. 8 Evolution of resistivity, RH and temperature along time.

Assessment of the residual life of the bridge

In order to make an assessment of the time to failure of the wires, the studies carried out were:

1) First, the designers of the bridge indicated a nominal load for the wires of 20 kN and they estimated that the ultimate limit state of the bridge will be

reached when a 55-65% of the wires of a tendon will fail in one of the four bridge piers.

2) In a separate study of the Materials Department of the Madrid Civil Engineering Faculty, they found no embritlement of this type of steel in alkaline solutions containing chlorides, on the other hand, they estimated the critical crack length K_{ISCC} of the wires. The rupture of the wires was characterized by the equation:

$$P_c[kN] = 45.0 - 11.4 \cdot a_c [mm]$$

From this equation considering the service load of 20 kN, the mean critical pit depth can be estimated of 2.15 mm.

With these previous considerations, the statistical study of the measured corrosion current was carried out.

Statistical treatment of the I_{corr} variability

Regarding the monitoring of I_{corr} values, two types of variations could be noticed: 1) that due to the climatic changes or the active-passive periods typical of pit evaluation , and 2) due to the differences in the position or localization of the testing points.

1) Concerning the effect of the climate (humidity and temperature), it has been already noticed (Andrade, 1996) in outdoor conditions, that seasonal changes induce a variation coefficient of about 1. This means that, assuming the Icorr is recorded at least during a continuous year, the values may vary from very low ones to a value of the double of the mean calculated by integration along the time. Therefore each testing point can be characterized by a "Representative" I_{corr}, I_{REP}, which represent the behaviour along time without considering any decay, as it was not noticed along the 3 years of testing.

2) On the other hand, the dispersion of the Icorr values when taking into account the several testing points, has to be treated differently; that is, regarding the structural performance. Due to the big scatter of results among the different points of measure, all the I_{corr} values were grouped in four variables corresponding to the piers of the bridge, assuming that the variations inside the group were random. In each group, the mean and the variance was estimated. The parameters of the common distribution functions adjusted to data were calculated by the method of moments. And the best fitted distribution was obtained for each pier by means of the χ^2 goodness of fit method, but taking into account the tails of the distributions.

The parameters of the I_{corr} measures ant the distribution function chosen, for each pier, are showed in table 1. There were small differences, in each pier, among the three types of distribution functions indicated. With the distribution adopted, the distribution of time to reach the critical pit depth in a wire can be calculated. A concentration factor between local and average attack of 10 has been adopted.

Table 1 Corrosion current values in each pier

Pier	Samples number	Mean value	Standard Deviation	Coefficient of Variation	Best fitted distribution
1	55	0. 12	0.11	0.91	Gamma
2	75	0.18	0.13	0.73	Weibull
3	73	0.19	0.20	1.08	Weibull
4	29	0.08	0.10	1.29	Lognormal

Assuming that the fraction of wire failures in a tendon of a pier, in a point in time, is equal to the probability of failure of a wire in this pier, at this particular time, an assessment of the mean value of the time to failure in each pier can be assessed, under the consideration that the failure of the pier happens when the failure of 55% of the wires of a tendon is reached, as indicated by the designers. That is: the failure will correspond to the time to reach the critical pit depth with the corrosion rate given by the 45 % percentile (fractile) of the Icor distribution function.

Table 2 Residual life of piers

	Pier 1	Pier 2	Pier 3	Pier 4
Characteristic Icorr, 95% fractile ($\mu A/cm^2$)	0.32	0.43	0.55	0.25
Maximum pit depth, 95% fractile ($\mu m/year$)	37	49	63	29
Characteristic residual life (5% failures) (years)	58	44	34	74
Icorr, 45% fractile ($\mu A/cm^2$)	0.08	0.13	0.11	0.04
Maximum pit depth, 45% fractile ($\mu m/year$)	9	15	13	5
Mean residual life (55% failures) (years)	228	139	163	428

The mean value assessed and a "characteristic" value of the residual life for each pier are shown in table 2. From this table can be seen that the worse piers are two and three. In these piers a time to failure can be foreseen of some one hundred years if no measures are taken and the corrosion follows similar rates.

Conclusions

The accidental presence of chlorides in the mixing water of a postensioned bridge has induced its premature additional reinforcement and its continuous monitoring in order to predict the time to failure of the tendons.

The on-site recording of the corrosion rate has shown a certain variability of the I_{corr} values along the time and in function of the location of the testing points. Therefore, averaging along time was made to characterize each testing point by a Representative I_{corr}, I_{REP}, value.

Regarding the time to failure, the structural performance was taken into account and therefore, a statistical treatment was made by pier. A period of time to failure of 55-65% of the tendons from 140 to 400 years was calculated. This long period enables a normal use of the bridge and to introduce further the additional prestressing foreseen.

Acknowledgements

The authors are grateful to the funding provided by the CDTI (Centro Técnológico para el Desarrollo Industrial) of Spain and by the consultant firm APIA XXI. They would like also to recognize the relevant contribution to the residual life assessment made by the Materials Science Dpt. of the Faculty of Civil Engineering of Madrid, by means of the study of the steel embritlement.

References

Andrade C., Sarría J. and Alonso C. *Statistical study of simultaneous monitoring of rebar corrosion rate and internal relative humidity in concrete structures exposed to the atmosphere.* 4th Int. Symposium on Corrosion of Reinforcement in Concrete Construction, Cambridge, June (1996).

EP-93. Instrucción para el Proyecto y Ejecución de Obras de Hormigón Pretensado, Madrid (1993).

Feliu S., González J.A., Feliu S. Jr and Andrade C. *Confinement of the electrical signal for in situ measurement of Polarization Resistance in Reinforced Concrete.* ACI Materials Journal, Sep-Oct., (1990) pp. 457-460.

Hausmann D.A. *Steel corrosion in concrete. How does it occur?.* Materials Protection, November, (1967) p. 19.

Rodríguez J., Ortega L.M., Casal J. and Diez J.M. *Corrosion of Reinforcement an Service Life of Concrete Structures.* 7th. Int. Conference on Durability of Building Materials and Components, Stockholm, June (1996).

Stern M. and Geary A.L. *Electrochemical Polarization, I: A Theoretical analysis of the shape of polarization curves.* J. Electrochem, Vol.104, No.1, (1957) p. 56.

Rehabilitation of a Damaged Reinforced Concrete Bridge in a Marine Environment

J.J. Carpio[1], T. Pérez-López[1], J. Genescà[2] and L. Martínez[3]

Abstract

Severe corrosion condition and the repairing works of reinforced concrete piles of a bridge in the marine tropical environment of the Gulf of Mexico have been studied. The bridge in service since 1982, is supported by sets of piles about 24 m length and a cross section of 0.45 m x 0.45 m, which were hammered down into sea soil. Early after construction, the concrete of the piles exhibited cracks up to 4 m length above sea level, which motivated major repairs at 5 and 11 years of service. The repair work done by a construction company consisted on the filling the cracks with a resin and encasing the piles with a steel mesh reinforced cement jacket in moulds of fiberglass. The corrosion condition of the piles was assesed after 11 years of service, by measuring half-cell potentials, electrical resistivity and chloride contamination of concrete. Half-cell potential and electrical resistivity measurements indicated that large portions of the piles are in favorable conditions for high corrosion activity of rebar. Chloride penetration was found to be significative according to chemical analysis on core samples of concrete drawn from the piles. The measured values of the first parameters are very similar for unrepaired and repaired piles. Concrete coring allowed to observe the corrosion state of the steel reinforcement. The results indicate that the steel is actively corroding in the unrepaired piles and migth be only slightly corroding in the repaired ones.

Introduction

Reinforced concrete structures are often exposed to marine environments where deterioration by corrosion of the steel reinforcement frequently occurs (1-6). It is well known that seawater damages concrete and

---

[1]Programa de Corrosión del Golfo de México, U.A.C., Campeche, México
[2]Facultad de Química U.N.A.M.,
[3]Instituto de Física U.N.A.M.. México, D.F.

43

corrodes rebar because of the presence of chlorides and sulphates, oxygen and other deleterious agents. Cracking of the concrete cover has a direct influence on the durability of reinforced concrete structures. In the presence of cracks, seawater easily reaches the reinforcement and eliminates the protective conditions of the oxide layer formed around steel in the alkaline environment of concrete. In these conditions, the corrosion of steel occurs and due to the increased volume of the corrosion products formed important mechanical stresses are induced to concrete, which contribute to the development of cracks, the spalling of the concrete cover and to accelerate the corrosion process.

Concrete bridge protection, repair and rehabilitation is a topic of big concern (7-8). Repairing consists normally on patching, overlaying or encasement of bridge components , in which the sound or chloride contaminated concrete has been left in place ; in this case, the rate of deterioration is dependent on the rate of corrosion of active corrosion sites and the initiation of new corrosion sites in critically chloride-contaminated areas. Rehabilitation of concrete bridge elements requires removal of spall areas that have been patched ; and removal , patching and overlaying or encasement   with low permeability concrete of the delaminated areas and all regions where the corrosion potential is more negative than -0.25 V/CSE. In this case. the cause of the corrosion process has been adressed since the deteriorated and critically contaminated concrete has been removed. Concrete repairs may include using materials such as portland cement concrete, latex modified concrete and polymer concretes. Repairing or replacement of reinforcing steel may also be included. Many techniques have been used to control corrosion of steel in reinforced concrete structures, including galvanizing, epoxy coating, inhibitors, cathodic protection and other electrochemical methods. These techniques offer various degrees of protection to the reinforcing steel in concrete, but each has its own problems and limitations in specific situations. Membranes and sealants applied on the concrete surface can reduce further penetration of chlorides ; however such systems are not impervious and they do have a finite lifetime before degradation occurs. Furthermore, under conditions where oxygen can penetrate into concrete, corrosion is still possible. The only safe way to inhibit corrosion is to provide the structure with a protection system that will work effectively when chloride ions are present in the viccinity of the steel reinforcement.

The objective of this paper is to present the evaluation of corrosion damage of reinforced concrete piles of a bridge exposed to the marine tropical environment of the Gulf of Mexico and to describe the behavior of the repairing procedure used on those piles. The corrosion condition of the piles was assesed after 11 years of service, by measuring half-cell potentials, electrical resistivity, chloride contamination of concrete, visual observation and others.

Statement of the Problem

The bridge "La Unidad" links the "Isla del Carmen" with mainland Campeche on the Gulf of Mexico in the Yucatán Peninsula (Figure 1). "La Unidad" is the second longest bridge (3.2 km, 2.0 miles) over seawater in Mexico and has been in service since 1982. The structure consists of 108 type I beam prestressed decks, 30 m (100 ft) span supported by a massive reinforced concrete block over 12 to 14 piles. The reinforced concrete piles are 24 to 30 m in length, with a 0.45 m x 0.45 m (1.5 ft x 1.5 ft) cross section. The steel reinforcement is contituted by 8 longitudinal bars 2.54 cm diameter and stirrups placed as close as every 15 cm in regions where the highest shear stresses are acting. The concrete mix of the piles used 500 Kg of OPC type I cement per cubic meter, had a water to cement ratio of 0.5 and although designed for reaching compressive strength of 250 Kg/cm^2, control specimens during construction showed that only values of about 230 Kg/cm^2 were obtained.

The piles are exposed to different environments : about one half of the whole length of pile is inside the marine soil constituting the foundation, a portion of about 8 to 10 m is exposed to the seawater, regions of approximately 1 m are exposed to tide and lengths of the piles from about 3 to 5 meters are exposed to the splash zone and the marine atmosphere.

Early after construction concrete cracking was observed on the piles and after 4 years corrosion of reinforcement was easily visible. This point was investigated and it was found that piles had been overhammered during founding into marine soil. Cyclic loading of the pile (service loads and seawater horizontal forces) contributed to widening and development of concrete cracks. Under these conditions, steel reinforcement started corroding and contributed also to crack development. After 11 years of service some cracks had reached about 4 m long from the top of the pile down to the seawater level and more than 1 cm width in some cases. The general crack pattern of the piles is shown in Figure 1. These problems motivated several minor repairs like patching and ocassionaly filling of the cracks with resins during the first years of service. Nevertheless, the intensive deterioration of the bridge as assessed by visual observations motivated major repairs of the piles.4

Repairing Procedure of Piles

In 1987 (after 5 years of service) one third of the total amount of piles was repaired. The same repairing technique was used in 1993 (after 11 years of service) to rehabilitate another third of the piles of the bridge substructure. The procedure of the repairing technique is described below.

Injection of the concrete cracks.- The cracks of the reinforced concrete piles were filled with an epoxy resin using common injection procedures. During this procedure, some cores in cracked regions were withdrawn for compressive strenght testing.

Steel mesh surrounding of piles.- A steel mesh (15 cm x 15 cm grid and 0.63 cm rod diameter) was placed around the concrete surface along the length of the pile exposed to the seawater, the splash and the atmospheric zone (from the sea soil level to the highest region of the pile).

Fiberglass encasement of the piles.- The piles with the surrounding steel mesh were covered from the sea soil level to the uppermost region with 2 m long U-shaped fiber-glass elements (0.5 cm thickness) that were tightly joined with an epoxy resin.

Cement mix placement.- A OPC type II cement mix without sand (water-cement ratio equal to 0.5) was poured (with a procedure that allowed seawater expulsion) inside the fiberglass encasement, covering the steel mesh and forming a cover 5 cm thick.

Thus, the repair work provides a thick and probably impervious physical barrier around the piles, which is formed by a reinforced cement cover and the fiberglass elements (see Figure 2).

Corrosion Inspection of Unrepaired Piles

In 1993, a group of unrepaired piles was inspected in order to evaluate the deterioration due to direct exposure to the aggressive marine environment for 11 years. The inspection techniques used were visual observation and determination of cracks pattern, half-cell potential mapping, measurement of electrical resistivity of concrete and determination of chloride concentrations in concrete (9-12).

Visual observation and cracks pattern.- Visual inspection was conducted to detect general defects and to established cracks pattern of the concrete ; both, the location and dimension (lenght and width) of the cracks were determined.

Half-cell potentials mapping.- Half-cell potentials with a copper-copper sulphate reference electrode (CSE) were measured on the concrete surface of the piles according to the ASTM C876-83 standard. The grid pattern for these measurements was as follows : three points in each face every 0.30 m along the entire length of the pile from its top to the seawater level (lower tidal zone). From these values isopotentials curves were determined and drawn.

Electrical resistivity measurements.- The four electrode technique used in soils was used to measure the electrical resistivity of concrete of the piles. The electrodes were located 3 cm apart and 0.5 cm in deep. Mesurements were undertaken on each face of the pile and at different levels starting from the seawater line.

Determination of chloride concentrations.- Concrete cores (10 cm in diameter) were taken out from different piles at various levels from seawater, in order to determine chloride concentrations at steel level. The concrete sample was crushed and pulverized to get particles passing no. 200 mesh. Pulverized samples were placed for 24 hours in water to extract free chlorides by a filtration procedure. The chemical analysis was done using Volhard's method.

Assessement of Repair Work Performance

The performance of the repair work undertaken in 1987 was also assessed in 1993. The corrosion conditions of the reinforcement were evaluated on two different group of piles with the same inspection techniques described in the previous section. The first group consisted on piles whose fiberglass encasement had been intentionally removed at that time, in order to assess possible deterioration of the steel mesh and the cement cover because of direct exposure to seawater, splash or brine. The second group consisted on repaired piles whose protective barrier (steel mesh, cement cover and fiberglass elements) was kept intact . In both cases, the partial or the complete barrier, was removed to allow direct inspection of the reinforced concrete piles.

Results

Inspection of unrepaired piles.

All the piles exhibited long and wide vertical cracks. These cracks have propagated along the portion of the piles above high tide seawater level, having several meters in extension and in many cases crossing the full section. Their width ranges from a few tenths of millimeter to about 2 cm. The results of visual inspection and measurements undertaken in one of the unrepaired piles are shown in Figure 3. The cracks pattern is drawn in the four faces of the piles superimposed to isopotential curves -0.20, -0.35, -0.50 and -0.60 V/CSE.

The -0.20 V/CSE curve is generally located in the highest region of piles. In this case the -0.35 V/CSE isopotential is located at 4 m from high tide level in faces A and B. The -0.50 and -0.60 V/CSE curves had significant variations in position in the piles. In the pile shown in Figure 3 the -0.50 V/CSE isopotential has its lowest position at 1.8 m level and its highest one at 2.8 m. The -0.60 V/CSE isopotential is generally located near the 1 m level, as in this pile, but in some cases it approaches the seawater level. In some of the piles, half-cell potentials are well below - 0.60 V/CSE near the low tide level, but below it, in inmersion conditions, the half-cell potential reached values as low as -0.70 V/CSE.

The inspected urepaired piles showed electrical resistivity values of concrete ranging from 2 kohms.cm near the sewater level to about 20-30 kohms.cm in the highest regions. In the pile shown in Figure 3, resistivity values of 30 and 2 kohms.cm were measured at two different locations at levels 4 m and 2.5 m, respectively.

In this same pile, the chloride contents in the concrete in contact with steel were determined on cores taken out at 4 m and 2.5 m levels ; the chloride concentrations were 0.4 and 2.9 % wt. of cement, respectively.

When concrete cores were extracted, severe corrosion of steel reinforcement was observed at 2.5 m level while at 4 m level the steel was protected.

Inspection of repaired piles.

Piles with reinforced cement cover (no fiberglass encasement).- Visual inspection of these piles showed severe damage of the cement cover where cracks and rust stains were observed (see Figure 4). This cement cover was very brittle, probably due to chemical deterioration because of direct exposure to seawater. Removal of the cement cover was easily achieved alllowing to observe that the steel mesh reinforcement was severely corroded and in regions near construction joints it had almost vanished. After removal of the steel mesh and the cement cover in , it was possible to measure half-cell potentials and electrical resistivity directly on the concrete surface of some regions of the piles.

The isopotential curves -0.20, 0.35 and 0.45 V/CSE obtained from one of the piles are drawn in Figure 5. It can be seen that half-cell potential values were well below -0.20 V/CSE in almost the whole length of the pile ; from about 2 m level down to seawater line potentials are more negative than -0.35 V/CSE and below 1.8 m level the potentials are lower than -0.45 V/CSE but never lower than -0.50 V/CSE.

Electrical resitivity values of concrete were 11.5 , 7.5 and 4 kohms.cm at 3, 2.5 and 1.6 m levels.

Piles with reinforced cement cover and fiberglass elements.- Visual inspection after 6 years of repair work allowed to observe that the reinforced cement cover and the fiberglass encasement were in good condition in most of the repaired piles, as seen in Figure 6. Fiberglass encasements were removed in some piles so that visual observation and measurements could be done on the surface of the original concrete.

During removal, no chemical attack of the reinforced cement cover nor corrosion of the steel mesh was observed and hammer testing proved the reinforced cement cover to be very resistant to impact. In some cases, physical damage, possible due to action of ultraviolet radiation, is observed in regions of the piles exposed to wetting and drying (tidal zone). Cracks on the fiberglass elements and the reinforced cement cover were observed in one of the piles in a region where the concrete was cracked before.

Half-cell potentials and electrical resistivity measurements were undertaken on two faces of the piles and chloride contents at steel level were determined on two concrete cores. The results of measurements are presented in Figures 7, where isopotential curves -0.35, -0.50 and -0.60 V/CSE were drawn. The position of the isopotential -0.35 V/CSE varies from 2 m to 3 m level and the -0.50 V/CSE curve goes from 1.30 m to 0.30 m level ; the -0.60 mV/CSE curve is located very close to the low tide level.

Electrical resitivity of concrete at levels 1.1 m and 1.75 m are 5.7 and 3 kohms.cm, respectively. At these same locations, chloride concentrations at steel level were 0.66 and 037 % by weigth of cement ; when concrete cores were drawn, the longitudinal reinforcing steel did not show evidences of corrosion.

Discussion

In the unrepaired piles, potentials measurements at and above sea level indicate that reinforcement along almost the whole length of the pile is probably corroding, assuming the criteria of the ASTM C876-83 standard (13). The case reported here shows that most of the pile presents potentials below -0.35 V/CSE and lower than -0.50 V/CSE below 3 m level, which means 95 % probability that steel corrodes. The same indications are drawn from the electrical resistivity measurements, as values as low as 5 and 2 kohms.cm were measured at 3 m level and near the water level, respectively. Chloride contents in concrete also indicate that corrosion of steel reinforcement is possible, as most of them are higher than 0.4 % by weigth of cement and in some cases they reach values as high as 3 % by weigth of cement. The corrosion conditions of the reinforcement in sites where concrete samples were cored agrees with the measurements forecastings (see Table 1).

On repaired piles with steel mesh, cement cover and fiberglass encasement, half-cell potentials were more negative than -0.35 V/CSE in a portion of the pile 2 to 3 m above low tide level. Electrical resistivity of concrete in that region was lower than 6 kohms.cm and chloride concentrations in contact with steel were very near the critical content for corrosion to occur (9) ; however, visual inspection after concrete coring show no evidence of corrosion of reinforcement.

Similar results were obtained on piles with only steel mesh reinforced cement mix. Potentials were well bellow -0.20 V/CSE in almost the whole length of the pile, and more negative than -0.35 V/CSE from about 2 m down to low tide level. Electrical resistivity values of concrete were 10, 7.5 and 5 kohms.cm at 3 m, 2.5 m and 1.6 m levels. These results would indicate that corrosion of reinforcement is very possible in a region of the pile going from the low tide level to 2 m level ; in the rest of the pile, corrosion conditions cannot be well determined as potentials ranged between -0.20 and -0.35 V/CSE. The electrical resistivity values are consistent with half cell potentials indications as values lower than 10 kohms.cm are found in almost the whole length of the pile and higher values are only recorded in the uppermost regions (see Table 2).

Conclusions

Piles that have remained unprotected during 11 years present very high chloride contents and the half-cell potentials and the electrical resistivity measurements indicate that corrosion of reinforcement is very likely to be occurring. Moreover, severe corrosion of steel reinforcement was found on these piles after 11 years of service. Although no corrosion was detected on piles repaired 6 years after construction., it should be noticed that chloride contents in concrete in contact with steel are very near the critical content, that moisture in the concrete is high (as can be stated from visual inpection and electrical resistivity measurements) and that dissolved oxygen can still be available in the pore water of the concrete, so that the corrosion of reinforcement could take place. The more aggressive environment (higher chloride contents and very wide cracks,) around steel in piles after 11 years of service let us think that the repairing procedure would be uneffective in slowering down the corrosion process.

References

1. N.J.M. Wilkins, P.F. Lawrence, "The Corrosion of Steel Reinforcements in Concrete Immersed in Seawater", Conf. on Corrosion of Reinforcements in Concrete Construction, London, 13-15 June 1993

2. M. Makita, "Marine Corrosion Behavior of Reinforced Concrete Exposed at Tokio Bay". Performance of Concrete in Marine Environments, ACI SP-65, Detroit (1980), pp. 271-290

3. J.J. Carpio, L. Martínez, T. Pérez López, "Severe Corrosion of a Gulf of Mexico Bridge". Materials Performance, December 1994, Vol. 33, No. 12, pp. 12-16

4. V. Novokshchenov, "Corrosion of Reinforced Concrete in the Persian Gulf Region", Materials Performance, Jaunuary 1995, Vol. 34, No. 1, pp. 51-54

5. O. de Rincón et al, "Electrochemical Diagnosis and Rehabilitation of Pilings for a Marine Bridge", Materials Performance, August 1996, Vol. 35, No. 8, pp. 14-21.

6. A. Sagües, R.G: Powers and R. Kessler, "Corrosion Processes and Field Performance of Epoxy-Coated Reinforcing Steel in Marine Substructures". Corrosion'94 NACE Conference, paper no. 299

7. SHRPC-S-360, "Concrete Bridge Protection, Repair and Rehabilitation relative to Reinforcement Corrosion : A Methods Application Manual, SHRP, National, Research Council, Washington, DC 1993

8. RP390-90 NACE Standard, " Maintenance and Rehabilitation Considerations for Corrosion Control of Existing Steel Reinforced Concrete Structures, NACE, Houston, TX, 1990

9. P.R. Vassie, "A Survey of Site Tests for the Assessement of Corrosion in Reinforced Concrete, TRRL Laboratory Report 953, Crowthorne, Berkshire (1980)

10. J.L. Dawson, "Corrosion Monitoring of Steel in Concrete" in Corrosion of Reinforcement in Concrete Construction, ed. A.P. Crane (Ellis Horwood Ltd., London: Soc. of Chem. Ind., 1983) pp. 175-191

11. S. Feliú, C. Andrade, "Manual de Inspección de Obras Dañadas por Corrosión de Armaduras", (Madrid, CSIC, 1992)

12. A. Raharinaivo, P. Brevet,. G. Grimaldi and J.J. Carpio, "Techniques for assessing the Residual of Lifetime of Reinforced Concrete Civil Works", 9th European Congress on Corr., paper no. BU097, The Netherlands, October 1989

13. ASTM C876-83, "Standard Test Method for Half-cell Potentials of Uncoated Reinforcing Steel in Concrete (Philadelphia, PA:ASTM)

REINFORCED CONCRETE STRUCTURES

Table 1 : Field measurements results for unrepaired piles

Core	Pile No.	Heigth[A] m (ft)	Potential[B] - Volt	Resistivity kohm cm	[Cl-] [C] % wt	Condition
T1	2.5	2.2 (7.3)	0.54	2.7	0.44	corroded
T2	2.2	3.3 (11)	0.38	3.0	0.88	corroded
T3	2.5	1.1 (3.6)	0.58	1.5	0.56	corroded
T4	1.2	2.5 (8.3)	0.46	2.0	2.92	corroded
T5	2.3	1.3 (4.3)	0.63	2.0	2.84	corroded
T6	1.6	1.0 (3.3)	0.57	4.0	0.48	corroded
T7	1.2	4.0 (13)	0.3	30.0	0.44	corroded
T8	2.3	4.2 (14)	0.18	12.0	0.07	noncorroded

(A) From sea level
(B) Electrical Half-cell potential of reinforcing steel in concrete, vs CSE. ASTM Standard C 876
(C) Chloride concentration in concrete cores, by weigth percentage of cement

Table 2 : Field measurements results for repaired piles

Core	Pile No.	Heigth, (A) m (ft)	Potential, (B) - Volt	Resistivity, Kohm cm	Cl- (C) Concentra- tion, %	Condition
T9	2.3	2.7 (9)	0.40	5.7	0.37	noncorroded
T10	2.3	1.8 (6)	0.46	3.0	0.66	noncorroded
T11	TEST	2.2 (7)	0.47	1.0	0.36	noncorroded

Figure 1: The bridge "La Unidad" links the "Isla del Carmen" with mainland Campeche on the Gulf of Mexico in the Yucatán Peninsula.

Figure 2: The repair work provides a thick and probably impervious physical barrier around the piles, which is formed by a reinforced cement cover and the fiberglass elements.

HEIGHT, m

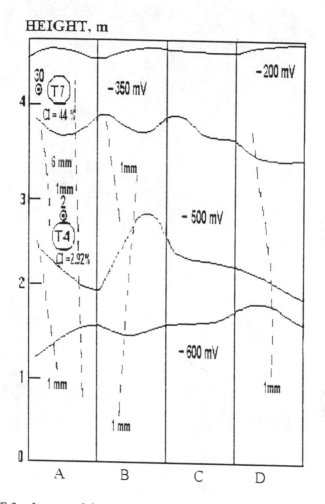

FIGURE 3 .- Isopotential curves, electrical resistivity measurements, chloride contents and corrosion condition of unrepaired pile.
Resistivity measurement indicated by ⊙ in KΩ-cm.

Figure 4: Pile with reinforced cement cover (no fiberglass encasement). Visual inspection of these piles showed severe damage of the cement cover where cracks and rust stains were observed.

HEIGHT,m

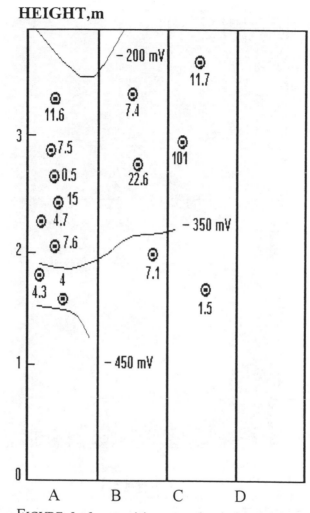

FIGURE 5.- Isopotential curves, electrical resistivity measurements, chloride contents and corrosion condition of repaired pile (cement cover protection and fiberglass encasement).
Resistivity measurement indicated by ⊙ in KΩ-cm.

Carpio, et al.

Figure 6: Piles with reinforced cement cover and fiberglass elements. Visual inspection after 6 years of repair work allowed to observe that the reinforced cement cover and the fiberglass encasement were in good condition in most of the repaired piles.

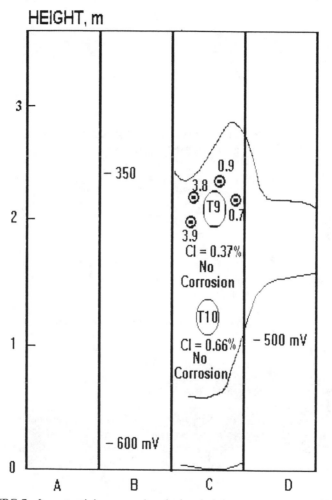

FIGURE 7.- Isopotential curves, electrical resistivity measurements, chloride contents and corrosion condition of repaired pile (cement cover protection, no fiberglass encasement).
Resistivity measurement indicated by ⊙ in KΩ in cm.

PGRU BRIDGE: REHABILITATION BASED ON AN ELECTROCHEMICAL DIAGNOSIS

Oladis T. de Rincón, Rafael Fernández, Matilde F. de Romero, Daniel Contreras, César Vezga, Oleyda Morón, Miguel Sánchez[1]

ABSTRACT

This paper presents a summary of the evaluation results, diagnosis and the repair/rehabilitation procedures for the General Rafael Urdaneta Bridge (PGRU) on Lake Maracaibo. Pier 9 was chosen because it is one of the most affected by corrosion in the reinforcement, both in the superstructure and the supporting piles.

The PGRU evaluation was as follows: general visual inspection, damage survey (delamination, porous concrete, cracks, etc.), measurement of potentials and reinforcement corrosion rate (determining Rp by confining the current), physical-chemical analysis and concrete resistivity. All these are complementary tests run when evaluating corrosion damage in structures such as these, providing the means for efficient repairs/rehabilitation to be carried out.

The survey shows that the damage is located on the edges, on which generally inadequate repairs have previously been carried out, as well as on the leeward faces of the shafts not exposed to the prevailing winds. Since they are subjected to greater humidity, it facilitates chloride-ion diffusion, thus causing faster reinforcement corrosion.

The inspection indicates that corrosion is basically due to chloride-ion ingress, particularly in areas with less protective concrete covering the reinforcement. It also shows that reinforcement corrosion does not actually affect the structural behavior of the bridge. The physical-chemical and electrochemical evaluation identified

---

[1] Universidad del Zulia. Centro de Estudios de Corrosión.Apdo. 10482.
Tel-Fax: 58-61-512197. .Maracaibo - Venezuela

the damaged areas and depths to which repairs must be made to guarantee effective results without affecting the areas with passivated steel. The repair contemplated removing the chloride-contaminated concrete with the Hydrojet technique, removing and replacing the corroded rebars and applying special mortars with the Shotcrete technique.

INTRODUCTION

The original design, materials quality control, and techniques and procedures used for the construction of the "General Rafael Urdaneta" Bridge (henceforth PGRU, Figure 1) made it one of the most successful engineering works of the Sixties. PGRU is supported by cables in the longest spans. At the time, it was the longest pre-stressed and post-tensioned concrete bridges in the world (8,678.6 m). It is located in northwestern Venezuela, facing the Caribbean, with four lanes measuring a total of 17.4 m. It has 135 piers of different designs and is supported by 2184 piles of three different types: hollow prestressed, hollow post-tensioned and reinforced-concrete piles.

Figure 1. View of the PGRU, Maracaibo, Venezuela

Studies for constructing PGRU were started in 1956. From the very beginning, reinforced concrete and not steel was contemplated because of the highly corrosive Lake environment (high oxygen concentration and high temperature), even though Cl⁻ concentration in its waters was 400 ppm at the time. Recent studies by Petróleos de Venezuela (PDVSA) have indicated that sea-water influx has increased corrosivity in Lake Maracaibo from 16 mpy/1962 to 180 mpy/1993, an increase in chloride ions from 400 to 3000 ppm, respectively. This effect, together with the high temperature and relative humidity, have exposed PGRU to a highly aggressive environment.

PGRU became operational in 1962, but by 1964 corrosion problems were detected not only in the reinforcement but also in the stays, all of which were replaced in 1980. Studies in 1981[9] determined that, in general, the main corroding factor was that the protective concrete layer on the reinforcement was too thin.

In 1984, an inspection by Rincón and Locke[13] revealed corrosion damage in the supporting piles of the first 19 piers. This led to studies and investigations oriented towards the application of cathodic protection (CP) for these piles. Two CP systems were designed for test piles: one using a bare Al-Zn-In alloy anode in the lake to protect the submerged area of the test pile; the other, using an anode bracelet of the same alloy embedded in Portland cement, to protect the tide and splash zones. In both systems, the steel potentials changed to levels consistent with those established for adequate cathodic protection[14]. An inspection in 1996 revealed that the steel reinforcement of the upper part of the pile was exposed, with active potentials values (<-250 mV vs Cu/CuSO$_4$); therefore, on the basis of the results obtained by Rincón and Locke[2], it was decided to install cathodic protection in one of these piles, using Al-Zn-In anodes directly submerged in the lake[15].

Now, 34 years later, the PGRU structure has deteriorated to such an extent that repair/rehabilitation must be addressed as quickly as possible. This paper presents the current diagnosis of the PGRU structure, using Pier 9 as a typical case study, since it is the one most affected by reinforcement corrosion in both the superstructure and supporting piles. The work is based on a physical-chemical, electrochemical and structural study that will enable adequate repairs/rehabilitation to be carried out, thus guaranteeing an extension of its service life.

DESCRIPTION OF PIER 9

Pier 9 (Figure 2), comprises a prestressed concrete platform supported by four crossed (X) shafts. These are of a variable rectangular section cast from the base of the caps with a 28 ϕ 20 mm main reinforcement; the bottom faces of the beams connecting the pile caps are at about 76 cm from the pile cap top, with a two-layer 10 ϕ 26 mm symmetrical main reinforcement on each face and a 6 ϕ 20 mm facing.

The upper and lower areas of the two caps are 4.20 x 15.60 m and 4.80 x 16.15 m, respectively. Total height is 3.00 m, including 30 cm of the prefabricated casing. These are joined by four 1.10 x 2.70 m connecting beams, including 70 cm of prefabricated concrete, cast with two 8 ϕ 26 mm layers each in both the top and bottom area, and three equidistant 4 ϕ 18 mm layers. The original indications stipulated 11 piles per cap. However, as some were rejected after inspection and for symmetry, 4 more were added per cap. These were driven between the 4 pairs of original piles in the central row, but shorter than the original ones. Total number of piles in the pier: 30.

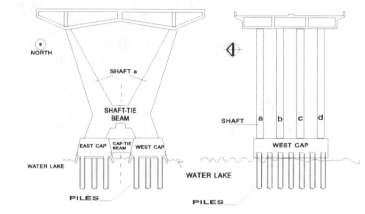

Figure 2. Pier 9: North and West faces description

EVALUATION: CRITERIA AND PROCEDURES

The following is a list of complementary techniques used to evaluate the pier: visual inspection, damage survey (delamination, cracks, etc), measurements: potentials, corrosion rate and resistivity, physical chemical analysis of the concrete, electrical continuity tests and depth of concrete cover.

Initially, the general inspection and damage survey was applied to the whole pier structure by detailed visual inspection, using a hammer to determine hollow-sounding areas, classifying the damage observed in accordance with ACI[12]. A Pacometer was later used to create a grid square showing the location of the steel in the areas where electrochemical measurements were to be taken. These were done to determine the amount of damage in the area indicated by the visual inspection. The Pacometer also enabled determination of concrete thickness over the reinforcement.

GECORR 6[4] was used to take the following electrochemical measurements: corrosion potentials (referring to a $Cu/CuSO_4$ electrode), reinforcement corrosion rate by linear polarization, concrete resistivity. This equipment also allows compensation of concrete resistivity and confines the current, thus enabling a definition of the affected area and the actual uniform corrosion rate. Before these measures were taken, a FLUKE multimeter was used to determine electrical continuity in the reinforcement. In the case of localized corrosion, it has been shown[3] that the pitting penetration rate is in the order of 10 times the value of the uniform corrosion value measured.

The physical-chemical evaluation was done by taking core samples at different locations, in accordance with the results obtained from the electrochemical evaluation. These samples (2" and 4" diameter and variable depth into the concrete) were dry-cut into 1.00 cm slices to determine chloride content with the MOHR method, thereby determining chloride-ion diffusion into the concrete and the approximate time they would take to reach the reinforcement (where electrochemical measurements indicate passivated steel). There was no other analysis because previous evaluations[8,9,10] had determined that PGRU steel corrosion is caused by chloride-ion ingress and that it has good quality concrete (compressive strength >300 kg./cm^2 and w/c ratio = 0.4).

AASHTO and COVENIN standards were applied to determine the structural behavior of the pier –as constructed and during repair– to determine the actual capacity ratio of each pier element, to enable repairs to be made without having to stop traffic.

Evaluation criteria

Potentials Measurements. Measurement of potentials by applying ASTM C876 can give an indication of potential reinforcement corrosion[2].

Corrosion Rate Measurements. It has been shown[6] that steel is passivated if the corrosion current density is <0.1 μA/cm^2; that corrosion is probable if the value is between 0.1 and 0.2 μA/cm^2; but that steel corrodes if this value is >0.2 μA/cm^2.

Resistivity Measurements. These help with the interpretation of results. It has been shown[1] that corrosion is very unlikely if resistivity is greater than 200 kΩ-cm, but that reinforcement corrosion is possible if it is very low.

Chloride Content. It has been fairly well established that reinforcement corrosion in marine environments (not immersed) is directly related to the level of chloride ions adjacent to the steel. It has been established[5] that this level should not be greater than 0.4% on the basis of cement content for reinforced concrete structures and 0.2% for prestressed concrete structures. Additionally, with the in-concrete chloride profile for older structures such as the one under study –34 years of age, under environmental conditions– it is possible to establish that the diffusional effect of the chlorides in the concrete is the most important, using Fick's 2nd Law to set up equations that could indicate the approximate time it would take for the chloride ions to reach the steel at levels that would cause depassivation[11]. Together with the electrochemical measurements, this measurement provides the means for performing an effective evaluation of the structure and executing better repairs/rehabilitation. They enable a definition of the areas where deep repairs must be made to the concrete without affecting the areas where the steel is passivated. In the latter, it would prevent further chloride-ion diffusion, thus giving a satisfactory extension to the service life of the structure.

Reduction in diameter of exposed steel. A reduction in the diameter of exposed steel in all areas with delaminated concrete enables evaluation of the effect it has upon each member's structural behavior. A diameter reduction of $\geq 10\%$ has been established[16] as the criterion for replacing the reinforcement.

RESULTS AND DISCUSSION

The whole pier structure was visually inspected and surveyed for damages, with the information being shown in plans for each pier element[16] (caps, shafts, cap-tie beams and piles). Figures 3 and 4 are photos of the pier showing the damage in the different elements.

Figure 3. Pier 9: North and West Faces. Greater amount of damages are shown on the West Face

Figure 4. Pier 9: South and East Faces. Greater amount of
damages are shown on the South Face with severe corrosion
of the reinforcement.

Note reinforcement corrosion in the areas nearest the lake and the more striking effect
observed in the Pier's south and west faces –the ones most protected from the
prevailing winds (northeasterly).This effect produces greater humidity on these faces,
inducing greater chloride-ion diffusion, which increases the reinforcement corrosion
rate.

**Detailed Inspection Results by area (shaft, shaft-tie and cap-tie beams, cap and
piles).**

SHAFTS AND SHAFT-TIE BEAMS. Even though Shaft "a" is exposed to the
prevailing winds (northeasterly), it is not the most deteriorated because it is directly
exposed to the sun, which reduces chloride-ion transport capacity towards the
reinforcement. There is localized damage only in the lower areas of the different faces
(given their proximity to lake splash), and very localized in some upper areas due to
thin concrete cover (<1.5 cm). Similar characteristics were found for Shafts "b" and
"c". The damage in Shaft "d", with similar characteristics to the ones just described, is
the most widespread because it is protected from the prevailing winds, thus facilitating
greater Cl⁻ diffusion into the concrete. Figure 4 shows the south face, the one with the
greatest damage in the lower area, and the damage survey for this Shaft is shown in
Figures 5. Figure 6 presents the results of the electrochemical evaluation of the same
face, which indicate reinforcement passivation.

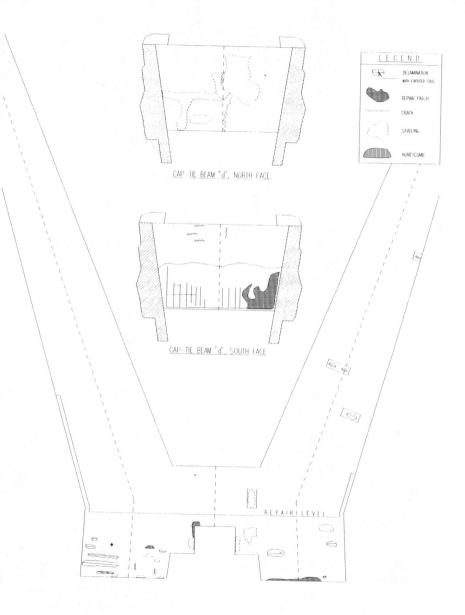

Figure 5. Damage Survey on Pier 9. South Face of the Shaft d.

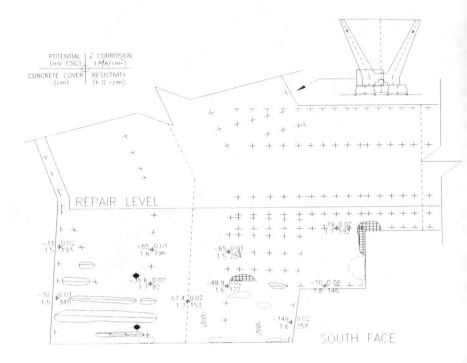

Figure 6. Electrochemical Evaluation on the South Face of Shaft "d'

However, note that the protective concrete layer is too thin (<1.5 cm) in some parts, so the reinforcement could become corroded in a short time, depending on chloride-ion ingress. These results, which are also typical of the other Shafts, provided the criteria for selecting the areas from which cores were to be taken for chloride-ion analysis.

Three samples of concrete were taken from Shaft "a" at the same height and in the areas with greatest damage (approximately 4.6 m above lake level and 1.32 m from the cap) to determine the chloride-ion content at different depths within the concrete. Results indicate that the chloride ions have not reached the reinforcement at concentrations that would cause corrosion (Figure 7). These tally with the electrochemical results (i_{corr} = 0,03 μA/cm^2 and E_{corr}= -34 mV vs Cu/CuSO$_4$). Fick's 2nd law shows that these ions would take more than 50 years to reach the reinforcement –generally found at depths of over 3.6 m (Figure 7 shows the D_{ap}

values). Given that Cl⁻ contamination has been found at less than 2.0 cm, however, it is expected that any reinforcement at less depth would be corroded –as was observed during the inspection.

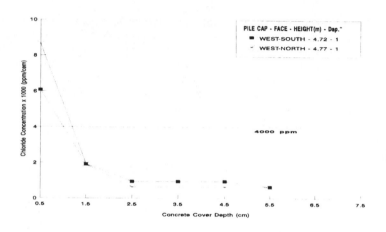

Figure 7. Pile 9. Chloride Profile in shaft "d".
Diffusion Coefficients (Dap)x10^{-09} (cm^2/s)

Five samples were taken from Shaft "d". Chloride-ion evaluation produced results similar to Shaft "a". These show that only the first cm of concrete contains this ion in sufficient quantities to induce reinforcement steel corrosion, with diffusion in the south face being greater than that in the north face (Figure 7). This coincides with the greater amount of damage indicated by the visual inspection. Similar results were found for Shafts "b" and "c". These results coincide with the electrochemical measurements, which determined that steel at greater depths than 1.5 cm was still passivated.

In accordance with the foregoing, the areas to be repaired would generally be restricted to the lower areas of the Shafts and in those localized areas where the steel is corroded because of too-thin concrete cover (Figure 5).As indicated by chloride-ion diffusion, repair depth will be restricted to 2 cm. However, 2.5 cm must be left clear below main rebars, so repair depth would be approximately 5 cm.

Further, delamination of the concrete, with moderate steel corrosion, can be seen inside all the orifices located at a height of 12 m on the two branches of each shaft, because of the thin concrete cover (<1.0 cm).

It is important to point out that all previously repaired areas show damages, with some faces being the only ones with reinforcement corrosion. This would not be the case if a detailed inspection had been carried out to obtain a proper diagnosis and work out an adequate solution for the corrosion problem detected.

CAP-TIE BEAMS. Since these elements are in the splash zone, they have suffered the most damage from reinforcement corrosion. The underside concrete is completely delaminated, with total or severe steel corrosion in many parts (Figure 6 and 8). All faces of the beams are either holow-sounding, delaminated or with evident signs of reinforcement corrosion (rust and/or cracking); here again, the south faces are the most affected. Electrochemical measurements indicate   additional reinforcement corrosion in those areas with no evident signs of damage yet.

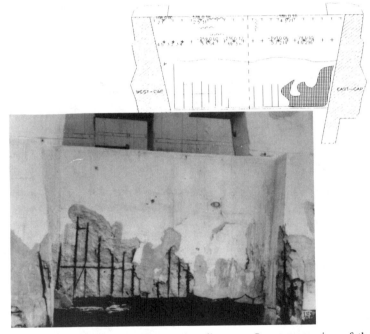

Figure 8.   Connecting   Beam "d" between the pile caps. Severe corrosion of the reinforcement and Electrochemical Evaluation on the South Face is observed.

Figure 9 shows some of the typical results of the chloride-ion test run on beams "a" and "d" (similar to the others). High chloride-ion diffusion can be seen in the concrete, with depths of 7 cm in the lower areas. These results tally with the ones above, thus proving that reinforcement corrosion is due to chloride-ion diffusion. An application of Fick's 2nd Law shows that the concentrations already at the reinforcement are enough to induce corrosion (3.3 cm). This coincides with the electrochemical measurements taken in this area (i_{corr} = 0.11 μA/cm^2, E = -248 mV vs. Cu/CuSO$_4$ and resistivity = 96 KΩ.cm). In this case, poor concrete quality (fς = 300 kg/cm^2 –determined for two different beams) allowed high chloride diffusion and subsequent reinforcement corrosion. The difference between shaft and beam concrete quality is notable (lower diffusion coefficient in the shafts –Figures 7 and 9).

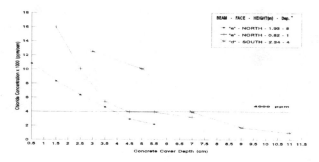

Figure 9. Pile 9. Chloride profile in the connection beam "a" and "b'between the pile caps. Diffusion Coefficients (Dap) x 10^{-09} (cm^2/s).

CAPS. As with the connecting beams, a high degree of reinforcement corrosion is generally evident in all the caps. Figures 3, 4 and 8 show that the top and all the other faces of these elements are cracked, with some hollow-sounding areas due to severe reinforcement corrosion. Damage is most severe on the corners/borders of the different cap faces, just as is the case with all the PGRU piers, where corrosion has reappeared in some areas because of inadequate repairs. Reinforcement corrosion in most cap/shaft corners, in spite of the very thick concrete cover in that area, was induced by inadequate compacting of the concrete, thus making it easier for the aggressive medium to penetrate. Corrosion is moderate to severe at different heights on the faces of both caps, with greater and more severe damages on the west and south faces, which are sheltered from the prevailing winds and the sun. Electrochemical measurements were taken in places with no evident signs of corrosion. The results indicate that the steel in some of these areas is already corroding.

The results of the chloride-ion analysis show high diffusion in the concrete: 5.5 cm deep on the windward and 6.5 on the leeward side, where diffusion is higher because of the environmental conditions described above (more humidity because it is sheltered from the sun).

The physical-chemical and electrochemical evaluation indicate the areas and depths at which repairs must be made to guarantee their effectiveness without jeopardizing any areas in which the steel is in excellent condition.

PILES. The visual inspection located evident signs of damage in only some prestressed piles, located among Piers 4-19, with severe reinforcement corrosion in some. Figure 10 shows the characteristic damages detected. In some cases, exposed rebars were completely corroded away.

Figure 10. PGRU Pile. Note the damaged concrete, with exposed steel. This pile is protected cathodically with submerged Al-Zn-In anodes. The probe used for the electrochemical measurements is shwon.

The failure of the concrete in most damaged piles was not due originally to reinforcement corrosion but to physical damage caused by wave action. When PGRU was being constructed, some piles were driven too deep into the lake bed or were so long that they had to be cut to design level (± 50 cm above lake level). In both cases a metal mold was used as shuttering (Figure 11) to level off the different piles and set up a prefabricated concrete casing that would then be used to erect the caps of the piles. The molds were left in place after additional concrete was poured into it to connect the pile with the pile cap. Wave and tidal action at the bottom of the metal mold increased turbulence in the space between the mold and the piles. This increased turbulence then eroded some or all of the concrete, enabling faster chloride ingress into the concrete to corrode the reinforcement. This type of damage was only found in several of the piles with metal molds.

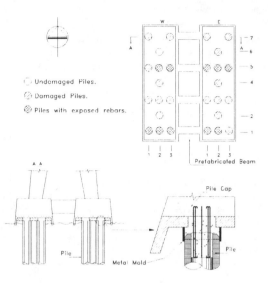

Undamaged Piles.
Damaged Piles.
Piles with exposed rebars.

Figure 11. Layout and damage survey of pilings in Pier No.9. Note how the metal mold was placed to level off the piles.

Additionally, there was evidence of a particular type of damage at the cap junction in some of these piles (Figure 12). As can be seen, when the concrete for the cap was cast, it did not completely fill the pile-cap junction area. This lack of compacting created voids, in some cases with severe corrosion in the steel joining the pile to the cap because there is no protective concrete in these areas.

Figure 12. Connection of pile and pile cap. Note the separation and the existing cavity with reinforcements corrosion

The survey (Figure 11) shows the damage of all the piles in Pier No. 9. Of these, 97% are deteriorated, 31% with exposed steel, it being the steel joining the pile to the cap in 44% of the cases. A thin copper flatbar was inserted into all piles with damaged cap junction to determine electric continuity between them, if there was exposed steel in the area, showing that the steel in some of these piles was already exposed.

Figure 13 shows the results of the electrochemical evaluation of one of the piles in Pier 9. They show that, even though the potentials are low (<-350 mV vs Cu/CuSO$_4$), i$_{corr}$ measurements are low (< 0.1 μA/cm^2). High corrosion rates were measured only at points adjacent to the area where the pile steel was exposed. This is due[15] to the galvanic effect produced by the exposed (corroding) steel and the passivated steel embedded in the pile. The status of the reinforcement connecting the piles to the cap was also evaluated. The electrical connection for these measurements was made with the cap steel, since it was shown (in those piles with exposed steel at the cap-pile junction), that there was a perfect electrical continuity where it joined the pile. The results (Figure 14) indicate localized reinforcement corrosion. There are other areas where measurement of potentials and current indicate passivity.

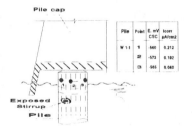

Figure   13.   Electrochemical evaluation of one of the PGRU piles

Figure   14.   Electrochemical evaluation of the reinforcement connecting the pile to the cap

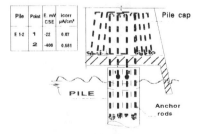

To evaluate chloride content in these elements, piles similar to those of the bridge were used (tested when PGRU was being constructed). Figure 15 shows chloride profiles at different heights and depths in different parts of the test pile. The effect of distance from lake level can clearly be seen in the chloride-ion contamination of the concrete, as shown by the very low level of these ions at 1.50 cm. However, in none of the cases have they got to the reinforcement level (prestressed concrete) in quantities large enough to cause depassivation (2000 ppm/cement). Fick's 2nd Law, applied to these data, indicates that it would take more than 50 years for the steel to become depassivated. The very low chloride diffusion coefficient values (1- 4 x 10^{-9} cm^2/s), indicate that the concrete in these piles is of excellent quality, which is in line with the specified design mixture (B-550 Portland Cement concrete), specific medium compressive strength = 531 kg/cm^2 with 500 kg concrete cement/m^3 and a w/c ratio = 0.36). These results may be extrapolated to the ones expected for the PGRU piles since these are exposed to similar conditions.

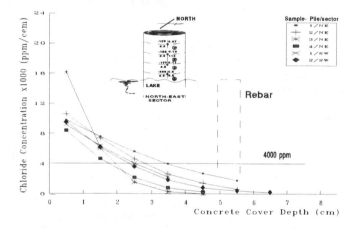

Figure 15. Pile 9. Chloride profile of a test pile

These results all indicate that only the pre-stressed PGRU piles with damaged concrete and exposed steel must be protected to avoid any further deterioration that would affect the rest of the pile, where the steel is passivated. The remaining piles have no reinforcement-corrosion problems and, therefore, additional protection is not warranted.

STRUCTURAL ANALYSIS

Using the survey and existing plans, a mathematical model was developed, with the help of the ISDS-STAAD III System. The weights of the 46.6 m stretches resting on the pier were calculated, plus the incidence of other permanent loads such as sidewalks and kerbs, protective rails, asphalt, etc. In this model, the caps were considered as formed by finite elements with variable dimensions according to the nodes chosen, with thickness being equal to its net thickness. Since no data were available with respect to the subsoil and the exact length of the piles, they were considered as articulated, vertically non-deformable supports.

Checks and investigation on trucks using PGRU have determined that loads of up to 90 tons on twelve wheelers. The rolling load was therefore calculated as a train equivalent to a 45-T truck per lane plus the impact and rolling-load reduction allowed in these cases by AASHTO standards. The result[16] is a total reaction per live load on the supporting axis of each 46.6-m beam. **In accordance with AASHTO, GROUP IA (1996) standards, this gives an amplifying factor of 2.08, as a function of the 45-Ton truck used.**

Brake load was taken as 5% with no impact, the same factor used in the original project. PGRU was not calculated for seismic movement, only for wind, considering 250 kg/m^2 for an empty bridge and 150 kg/m^2 when loaded, loads much smaller than the seismic loads contemplated in current COVENIN standards for Zone 2, Type I, S2 soil and works for public use, as is the case with PGRU. Due to the lack of antiseismic casting details, an elastoplastic spectrum was chosen for dynamic analysis with values that would not produce strains and stresses to put the structure in danger of collapse. This was done by reducing the spectrum until the greatest strains and stresses on the shafts fell within the volume of interaction, supposing that the shafts would be the first to collapse. Thus, the design (verification) spectrum was the one indicated[16]. The pier is rather flexible, its fundamental vibration period is 4.3 s in the direction of the bridge's axis, the second mode is 3.4 s across the bridge. The first six vibration modes were analyzed, the last one being 0.4 s. All of this indicates that for an elastic spectral acceleration of 0.322g, a ductility factor of 2 can be considered, and a maximum acceleration of 0.161g, with the **earthquake magnitude being 7.37 on the Richter scale, with expanding speed and displacement values of 19.53 cm/s and 11.73 m, respectively, for a focal distance of 67.75 Km.**

The same analysis was used to carry out the structural analysis for repairing the pier[16], considering: a) The cap-tie beams as though the lower part of the prefabricated part does not exist; b) The shafts and their anchoring beams with a smaller section, a product of the removal of 6 cm around the perimeter, as with the cap. The behavior of the pier is almost identical to that of the model as constructed[16]. All of this helped to set up a repair procedure that avoided having to shore up the pier and close the bridge to traffic.

Verification of the shaft sections at the foot of the Shaft-Tie Beams and at the table beam junction reveals that these sections fail with the chosen spectrum of 0.161g because the deadweight load conditions plus or minus earthquake fall outside the interaction surface. The design spectrum must therefore be reduced even more. **In order to avoid shaft failure in these places, the design spectrum must definitely be 0.10g. This means that the maximum earthquake that this pier can withstand is 6.77 on the Richter scale, with expanding speed and displacement values of 10.77 cm/s and 5.74 m, respectively.**

The structural study indicates that the pier can be safely repaired by strips in accordance with the specifications given in the following repair protocol.

GENERAL REPAIR PROTOCOL

Repairs will be carried out element by element, in the following sequence: Shafts and their Shaft-Tie Beams, Cap-tie beams, Caps and Piles.

Except for the piles, the following activities shall carried out on each of the elements indicated above:

1.   Removing concrete in areas with exposed rebars, in accordance with conventional procedures[7].

2.   Substituting exposed rebars in accordance with Welding Procedure Specifications[18].

3.   Removing concrete by the Hydrojet technique.

4.   Inspecting rebars and ties to determine diameter reduction.

5.   Preparing the steel surface by Specification SSP-2.

6.   Casting and curing mortar and/or concrete, as specified in Table 1, by the Shotcrete technique and according to Standard ACI-506.

7.   Applying Alkoxysilanes-based hydrofugous coating[18].

Table 1 .Proportions of mixtures used to repair the PGRU piers.

MIXTURE MATERIALS	MORTAR	CONCRETE
Water / Cement	0.40	0.40
Type 1 Portland Cement (kg.)	775	502
Coarse Aggregate (T.M.N.3/8") (kg.)	---	899
Fine Aggregate (3/8" Trickle) (kg.)	1102	704
Microsilica (% cement weight)	15	15
Polypropylene (Fiber) (kg/m³)	1	1
Superplastifier (% cement weight)	2	2.5

Tables 2 and 3 show some specific details for each of the elements scheduled for repair/rehabilitation.

Table 2 . Specific Details for Shaft, Shaft-Tie Beam and Cap Repair/Rehabilitation

		REMOVING DETERIORATED/CONTAMINATED MATERIAL	REPLACING REINFORCEMENT/CONCRETE/ MORTAR
S T R U C T U R A L	SHAFTS & SHAFT-TIE BEAMS	• General procedure: In large areas, concrete shall be removed in 75 cm-high strips[18], 5-cm deep, using the Hydroject technique; high-frequency percussion equipment shall be used in small areas, with prior delimitation of area to be removed with a ceramics cutting disk • Metal bars shall be cut at 45°. Bar length shall be delimited by the electrochemical measurements • Concrete in structural cracks shall be removed with high-frequency electric percussion equipment: 2.5 cm behind bars, 10 cm wide and the lengths indicated in the specific plans[16], making 2.5 cm "V" openings for resin injection	• New bar sections shall be replaced by welding to produce a butted V- joint[18] with filler material. Once replaced, the void space shall be filled with mortar. Wait 24 hours before another replacement in the same Shaft. • In large areas, use Shotcrete to replace concrete with special mortar[18] (Table 1). In small areas, repairs shall be conventional, using a polymer adhesive before applying the mortar. • Structural cracks shall be repaired by injection, using Standard ACI-224.1R.

| ELEMENTS | CAPS | • As a general procedure, concrete shall be removed up to a depth of 7 cm, using the Hydrojet technique.

• Remove concrete from the side faces, leaving a 30-35 cm strip on the upper part of the cap, which shall be removed together with the cap top.

• Metal bars shall be cut at 45°. The length of the bar to be replaced shall be delimited by the electrochemical measurements. | • Fit new bar sections into place producing a butted V-joint[18] with filler material.

• The cap top bars shall be welded leaving an overlap 10 times the diameter of the bar.

• The underside of the cap shall be cast in place to a height of approximately 1 m.

• A minimum 7-cm cover shall be guaranteed for the reinforcement. |

Table 3
Specific Details for Cap-tie Beams and Pile Repair/Rehabilitation

		REMOVING DETERIORATED/CONTAMINATED MATERIAL	REPLACING REINFORCEMENT/CONCRETE/MORTAR
STRUCTURAL ELEMENTS	CAP-TIE BEAMS	• High-frequency percussion equipment shall be used to remove the contaminated concrete on the periphery up to 20-cm at the bottom of the beam and 10 cm on the other faces. • The Hydrojet technique shall be used to remove the last 5 cm of each face. If the concrete nucleus is fractured, it shall be totally demolished. • Metal bars shall be cut at 45° with an abrasive disc. Before cutting, all corroded steel shall be removed until sound steel is found anchoring the beam to the cap.	• All steel bars shall be replaced, using two sections reinforcement. • Longitudinal bars shall be fastened by U-links, with half-beam overlap welding. • Concrete shall be cast into the molds; if only part of the beam is to be reconstructed, the concrete shall be cast into the lower end up to a minimum height of 40 cm, using the Shotcrete technique for the remainder. Minimum concrete cover shall be 7 cm.
	PILES	• Pile rehabilitation shall be done by applying cathodic protection with Al/Zn/In sacrificial anodes[15,17]. • The reinforcement in the Pile-Cap joint shall be protected with a bracelet-type anode embedded in Portland cement[15,17].	

Acknowledgment: This Detailed Inspection was made possible by a PGRU Repair/ Rehabilitation Agreement between the Government of the State of Zulia and 'Universidad del Zulia'.

REFERENCES

1. Alonso C., Andrade, C., González, J.A.: **Relation Between Resistivity and Corrosion Rate of Reinforcement in Carbonated Mortars Made with Several Cement Types.** Cement and Concrete Research., vol. 18, (1988) p. 687.

2. ASTM. Standard C-876-87.: **Standard Test Method for Half-Cell Potentials of Uncoated Reinforcing Steel in Concrete.** Philadelphia, P.A. ASTM. 1987.

3. Feliú, J.A. González; M.L. Escudero, S. Felín Jr., C. Andrade. **Corrosion 46** (1990), p.1015.

4. Feliú, J.A. González; V. Feliú; S. Jr.; M.L. Escudero; I.Rz. Maribona; V. Austiín, C. Andrade; J.A. Bolaño; F.Jimenez.: **Corrosion-Detecting Probes for use with a Corrosion-Rate Meter for Electrochemically Determining The Corrosion Rate of Reinforced Concrete Structures.** U.S. Patent, No. 5.259.944 (1993)

5. Funahashi, **ACI Materials Journal,** Nov-Dic.(1990): p-581.

6. González, J.A., Rodriguez, P., Feliú, S.: **Steel Corrosion Rates which Start to Introduce Problems in Reinforced Concrete Structures.** 1er. Simposio Mexicano en Corrosión, (1994), Mérida, México.

7. ICRI Technical GUIDELINES. Guideline No. 03730- **Guide for Surface Preparation for the Repair of Deteriorated Concrete Resulting from Reinforcing Steel Corrosion** (1995).

8. IMME (Instituto de Materiales y Modelos Estructurales) Universidad Central de Venezuela. **Puente General Rafael Urdaneta sobre el Lago de Maracaibo Informe de una evaluación previa de la Corrosión.** August (1971).

9. INGESA/LUZ Report: **Análisis e Informe sobre los Daños presentados en los Pilotes del Puente General Rafael Urdaneta sobre el Lago de Maracaibo. Estado Zulia.** Presented to MTC. December 1981.

10.Ingeniería Villanueva, S.R.L. **Prevención y Tratamiento de la Corrosión en el Puente General Rafael Urdaneta.** January (1975).

11.Poulsen, Ervin. **On a model of Chloride ingress into concrete having unidependent diffusion coefficient**, Nordic Miniseminar. Chaimers University of technology, Goteborg Sweden (1993).

12.Tibor Javor.: **Draft Recommendation for damage classification of Concrete Structures**. Materials and Structures, 1994, 27, 362-369.

13.Troconis de Rincón, O., Locke, C.: **Sacrificial Anodes - An Alternate Method to Protect the Reinforcing Steel in Concrete**. CORROSION/85, paper 260, (Houston, Tx: NACE International, 1985).

14.Trocónis de Rincón, O., Romero de Carruyo, A., García, O.: **Protección Catódica por Anodos de Sacrificio de Estructuras de Concreto Reforzado**. Revista Técnica de Facultad de Ingeniería. Special Edition. 10, 1 (1987). Venezuela.

15.Troconis de Rincón, O., Fernández de Romero, M., Contreras, D., Morón, O., Ludovic, J. and Bravo, J.: **Electrochemical Diagnosis and Rehabilitation of Pilings for a Marine Bridge**. Materials Performance, vol. 35, No. 8 (1996). NACE International, USA. pp.14-21.

16.Trocónis de Rincón, O., Fernández de Romero, M., Contreras, D., Fernández, R., Vezga, C., Morón, O., Bravo, J., Ludovic, J., García, O.: **Puente sobre el Lago de Maracaibo General Rafael Urdaneta: Evaluación, Diagnóstico and Reparación de la Pila 9**. Report presented to Obras Públicas del Estado Zulia (1996). Maracaibo, Venezuela.

17. Trocónis de Rincón O, Fernández de Romero, M., Contreras, D., Vezga, C., Bravo, J.: **Pilotes del Puente General Rafael Urdaneta: Diagnóstico Electroquímico and Rehabilitación**. Report presented to Obras Públicas del Estado Zulia. (1996). Maracaibo, Venezuela.

18.Trocónis de Rincón, O., Fernández de Romero, M., Contreras, D., Fernández, R., Vezga, C.: **Protocolo General de Reparación de las Pilas del PGRU**. Report presented to Obras Públicas del Estado Zulia. (1997). Maracaibo, Venezuela.

Condition Assessment and Deterioration Rate Projection for Chloride Contaminated Reinforced Concrete Structures

William H. Hartt[1], S. K. Lee[1] and Jorge E. Costa[2]

Abstract

A protocol for predicting the rate of chloride induced corrosion distress to existing reinforced concrete structures is projected by considering that time-to-failure as being statistically distributed and comprised of three component periods: 1) the time for corrosion initiation (TTC), 2) the time subsequent to corrosion initiation for concrete surface cracks to appear (Period II), and 3) the time for surface cracks to develop to a limit state or failure condition (period III). This evaluation method was applied using field survey data and laboratory test results for the Long Key Bridge in southern Florida which has experienced extensive chloride induced corrosion damage to the reinforced concrete substructure. Here, average values for the surface chloride concentration and the threshold were employed in conjunction with probability of occurrence data for the diffusion constant and reinforcement cover to calculate a distribution for TTC. The duration of Period II was taken from a literature projected value; and Period III was estimated from, first, definition of three possible substructure failure modes (limit states) and, second, estimation of when these will be reached considering present condition of the most deteriorated substructure components. From these the distribution for the net time-to-failure was projected. This, in turn, facilitated development of an informed plan for future repair and rehabilitation.

Introduction

It is now generally recognized that reinforced concrete structures exposed to chlorides, typically either from deicing salts in the case of northern climates or sea water for marine ones (or both), experience corrosion induced deterioration and abbreviated service life compared to situations where this species (chloride) is absent. Thus, while embedded steel in concrete is normally passive and corrosion rate accordingly low,

---

[1]Center for Marine Materials, Department of Ocean Engineering, Florida Atlantic University, Boca Raton, Florida 33431

[2]Corrosion Restoration Technologies, Inc., 612 N Orange Avenue, Suite A-11 Jupiter, Florida 33458

accumulation of chlorides at the steel depth in a critical amount compromises the protective film and, in the conjoint presence of moisture and oxygen, induces active corrosion. The resultant solid corrosion products accumulate in the concrete pore structure immediate to the steel-concrete interface and induce tensile stresses in the latter material (concrete). Because concrete is relatively weak in tension, cracking and spalling ultimately follow. Depending upon the type of structure and its service function, the reduced concrete section or continued reinforcing steel corrosion, or both, eventually compromise load bearing capacity to the point where the limit state (end of useful life) is reached.

An important aspect of the management and maintenance functions of reinforced concrete structures which are subject to chloride induced corrosion and deterioration is projection of, first, the present condition and, second, the rate at which continued deterioration can be expected to occur. Only from this information can an optimized decision process be developed for repair, rehabilitation, and, ultimately, partial or full replacement. In this regard, one recent report has addressed a condition evaluation procedure (Cady and Gannon, 1992) and a second developed the basis for a life-cycle cost analysis protocol (Purvis et al., 1994).

Protocol for Deterioration Rate Determination

The critical parameters which contribute to chloride induced deterioration of reinforced concrete structures are 1) chloride concentration of the environment (this determines the concentration of this species at the concrete surface which, in turn, influences the rate of its inward migration), 2) extent of wet-dry cycling (this contributes to chloride, water, and oxygen ingress in association with capillary flow and, hence, availability of these species to promote corrosion), 3) depth of embedded steel, 4) chloride diffusion constant for the concrete, and 5) critical chloride concentration for passive film breakdown.

An important aspect of any analysis based upon the five parameters listed above is that each is statistically distributed such that a probabilistic based approach is appropriate. However, parameters 1) and 2) are environment or exposure related such that data pertaining to these can normally be averaged over a time period which is much shorter than the remaining life of a particular structure to yield a specific value. In this regard, Bamforth (1993) reported that although surface chloride concentration may be a function of climatic conditions (rainfall, for example) and location on the structure relative to the chloride source, otherwise this concentration remains relatively invariant with time subsequent to the initial several weeks of exposure. On the other hand, parameters 3-5 are structure/material based and, as such, are spatially variable with changes with time being less significant. However, information regarding the last of these parameters (chloride threshold) is limited to the extent that some uncertainty exists regarding even an average value, much less how it might be statistically distributed.

In a generalized sense, deterioration of reinforced concrete structures as a consequence of exposure to chlorides can be represented in terms of three component periods: 1) time to corrosion initiation (TTC), 2) time for surface cracks to appear periods: 1) time to corrosion initiation (TTC), 2) time for surface cracks to appear subsequent to corrosion initiation, and 3) time for surface cracking to evolve to a limit state condition, as illustrated schematically in Figure 1. The first of these times (TTC) has been evaluated previously (Weyers, 1993) in terms of Fick's second law,

$$\frac{\partial c}{\partial t} = \frac{\partial}{\partial x}\left[D\frac{\partial c}{\partial x} \right],$$

1)

where

$\partial c / \partial t$ is the change in concentration of the diffusing species at a location of interest with time,

D is the diffusion constant,

c is concentration of the diffusing species, and

x is distance,

and its solution, assuming that D is independent of c,

$$c(x,t) = c_o\left\{ 1 - \mathrm{erf}\left(\frac{x}{2(D \cdot t)^{1/2}} \right) \right\},$$

2)

by setting $c(x,t)$ equal to c_{crit}, the critical chloride concentration threshold and x equal to the reinforcement cover and solving for t as TTC. In the present analysis Equation 2 was extended to a probabilistic format as

$$c'(x,t) = c_o\left\{ 1 - \mathrm{erf}\left(\frac{x(p')}{2\big(D(p'') \cdot t\big)^{1/2}} \right) \right\},$$

3)

where $x(p')$ and $D(p'')$ are the concrete cover and diffusion constant, respectively, corresponding to a particular probability of occurrence (the probability of occurrence pertaining to one of these parameters need not be the probability for the other) and $c'(x,t)$ is the resultant chloride concentration. The problem then reduces to setting $c'(x,t) = c_{crit}$ and solving for t (TTC) in terms of various $x(p')$ and $D(p'')$ combinations, where data pertaining to the latter two parameters are arrived at from statistically significant x and D determinations for the structure of interest. From the resultant $x(p'):D(p'')$ versus TTC trend, the rate of future corrosion initiation can be projected.

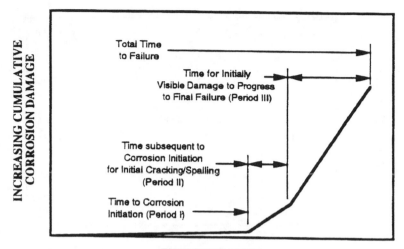

EXPOSURE TIME

Figure 1: Schematic representation of the three regimes of corrosion damage
leading to final failure.

There is presently no analysis protocol for predicting the time for surface cracking to
appear subsequent to corrosion initiation (Period II in Figure 1); however, a time of 3.5
years has been reported based upon past field studies (Weyers, 1993). On the other hand,
projection of Period III (time for surface cracking to develop to the limit state condition) is
likely to be structure specific.

Example Protocol Application - The Long Key Bridge

General

The Long Key Bridge is located along the southern tip of the Florida peninsula and
constitutes a portion of the overseas highway to Key West. It was constructed in 1980,
consists of 103 double vee piers which are equally spaced along its approximately 3,380 m
length, and extends over essentially open sea water. Because of the corrosive nature of the
exposure, epoxy coated reinforcing steel (ECR) was employed in the construction of all
substructure components; however, initial indications of corrosion induced concrete
cracking and spalling were apparent here as early as 1986 (Kessler and Powers, 1986);
and the incidents of these damage occurrences has progressively increased since that time,
as illustrated by Figure 2 (Sagues et al., 1995). No corrosion related problems are evident
upon the superstructure, however, which is consistent with the Florida experience that
damage occurs primarily within the 0-4 m elevation range above mean sea level. As a
consequence, the present investigation focused solely upon the substructure.

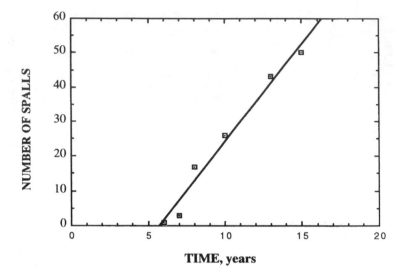

Figure 2: Number of spalls detected upon the Long Key Bridge
substructure as a function of time.

Figure 3 is a photograph and Figure 4 provides a schematic illustration of a typical
double vee pier and shows these to consist of 1) two 1.07 m diameter drilled shafts, 2) a
cast-in-place footer cap upon each of the two drilled shafts with an interconnecting, precast
strut, 3) a precast pile cap which spans the footers and rests upon a reinforced neoprene
pad centered upon each of the two footers, and 4) four precast legs which extend to and
support the segmented box beam superstructure. The design of the new bridge was such
that mean sea level was approximately 0.3 m below the base of the footers. As such the
drilled shafts are expected to be water saturated; and, consistent with this, no corrosion
related damage upon these has been noted. Consequently, this evaluation focused upon the
footer/struts, pile caps, and legs only. More detailed features of the substructure include 1)
a plastic hinge near the top and bottom of the legs of every sixth bent to accommodate
longitudinal movements, 2) a shear key on the underside of the pile cap at mid-span of the
bents to either side of the hinged leg ones to restrict lateral movements, and 3) a local mat
of relatively dense reinforcing steel (13 mm diameter bars with 100 mm spacing) near the
top footer surface beneath the bearing pad and near the underside of the pile cap just above
the pads. As such, the latter two component types (footers and pile caps) contain two types
of reinforcement detail; a cage and a bearing pad grid, whereas the others (struts and legs)
contain cage type of reinforcement only. The cover over the reinforcement cage in all
cases was specified as 100 mm. However, the bearing pad grit was placed outside the
reinforcement cage; and, thus, cover for this steel was less. Also, the key at the base of
certain pile caps (see above) is heavily reinforced with a specified cover of 50 mm.

Figure 3: Photograph of a typical Long Key Bridge pier.

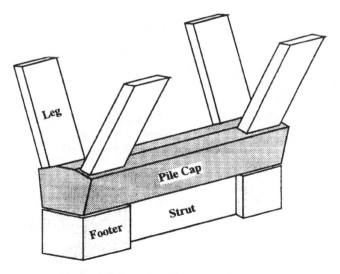

Figure 4: Schematic of Long Key Bridge pier.

The investigation was comprised of both field and laboratory studies, where the former consisted of visual damage characterization, acquisition of cores, and measurement of cover over the ECR, while the latter involved analysis of the cores petrographically and for chlorides and of ECR samples contained within some of the cores. The purpose of these studies was to 1) define the present condition of the substructure from a corrosion deterioration standpoint, and 2) provide input information for the damage accumulation model (Equation 3) so that a plan for future bridge substructure maintenance (repair, rehabilitation and/or replacement) could ultimately be made.

The laboratory analysis of the ECR samples involved characterization of both the bars and the epoxy-coating. This consisted of visual inspection for corrosion damage, electrochemical impedance spectroscopy (EIS) measurements, coating thickness determinations, and documentation of the type and density of coating defects. The results of these measurements, as they pertain to condition of the coating, are not addressed here other than to point out that, first, the coating quality conformed to the relevant specification which existed at the time of construction with the exception that excessive coating defects were present and, second, the epoxy coating had not provided the degree of protection to the embedded steel that has normally been projected to result for this corrosion control method. This latter point is consistent with previous inspections and analyses which concluded that onset of ECR corrosion in this structure was comparable with what was expected for bare reinforcement (Kessler and Powers, 1987). In the analysis consideration was given to the type of component involved (footer/strut, pile cap, and leg) since the design, construction, exposure, and structural function differs for each.

Visual Damage

Results of the visual inspection revealed that cracking and spalling were most evident in association with 1) the lower edge bars of the pile cap reinforcing cage, 2) at the keys (underside of certain pile caps), and 3) the bearing pad grid of both the footers and pile caps. No corrosion induced cracking and spalling of the legs and only two instances for the footer reinforcement cage were disclosed.

Data Development

Chloride Analyses. A total of 65 cores (100 mm diameter) were acquired from the various substructure elements (see above), and powdered concrete samples were obtained from all of these either by drilling with a masonry bit or by dry slicing parallel to the exposed surface with a diamond blade. One such sample was taken at the 50 mm depth for all cores and at multiple depths in 24. Figure 5 presents typical chloride concentration (ppm Cl^- on a concrete weight percent basis) versus depth data as acquired from a leg element. These reveal that the acid soluble concentrations were only slightly greater than the water soluble, thereby suggesting that a relatively
small percentage of the chlorides was chemically bound. All subsequent analyses were by the acid soluble technique.

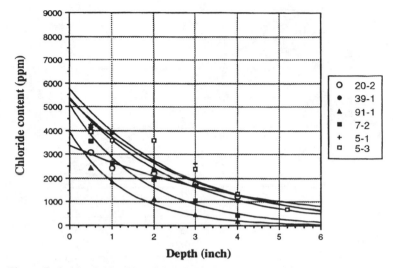

Figure 7: Acid soluble chloride profiles of the cores taken from Pile Caps.

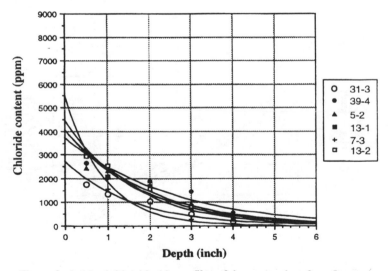

Figure 8: Acid soluble chloride profiles of the cores taken from Legs.

Results of a statistical analysis of these data sets are shown in Figures 9-11, respectively, as chloride profiles corresponding to 97.5, 90, 80, 70, and 50 percent probabilities of occurrence.

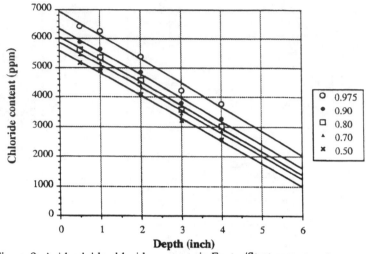

Figure 9: Acid soluble chloride contents in Footer/Strut component
with different confidence intervals for μ.

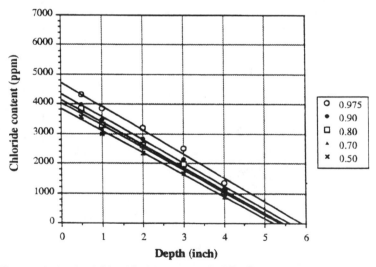

Figure 10: Acid soluble chloride contents in Pile Cap component
with different confidence intervals for μ.

Thus, the 97.5 percent probability of occurrence data can be interpreted to mean that three pile caps, six footer/struts, and 18 legs of the 103 bents should have chloride concentrations which equal or exceed this limit.

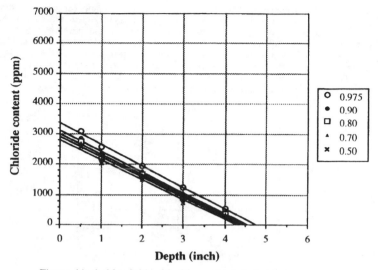

Figure 11: Acid soluble chloride contents in Leg component
with different confidence intervals for μ.

Figure 12 shows the same chloride analysis results at the 50 mm depth from Figures 6-8 in a bar graph format along with the 50 mm cover depth data for the remaining cores. Listed also is the mean 50 mm depth chloride concentration value for each case. Comparison of the mean values from Figures 9-11 at the 50 mm depth with the corresponding values from Figure 12 indicates reasonable mutual agreement.

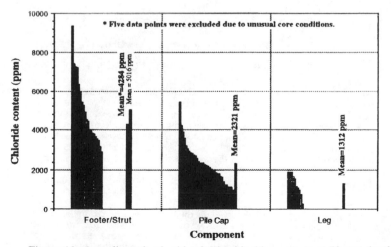

Figure 12: Overall result of acid soluble chloride contents at 50 mm depth.

The scatter for the data in Figure 12 for a particular component indicates that concrete quality was variable either with location upon a particular member or from one pier to the next (or both). With regard to the former possibility, the fact that cores were not taken at the same location on all components of the same type could have contributed to the scatter. To determine if this was a factor, an attempt was made to correlate chloride concentration with elevation relative to the base of each of the three component types (see Figure 4). Figures 13-15 show schematic elevation views of the footer/strut, pile cap, and leg components, respectively, and identify coring locations upon each.

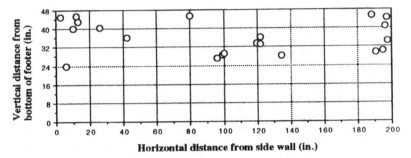

Figure 13: Core extraction locations in footer/strut component.

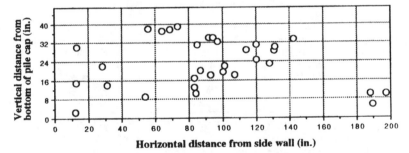

Figure 14: Core extraction locations in pile cap component.

Correspondingly, Figure 16 presents the 50 mm chloride concentration data from each core on the respective structural component as a function of elevation on that component. These results indicate that, while chloride concentration for a particular component type (footer/strut, pile cap, or leg) did vary in inverse proportion to its elevation above sea level (see Figure 4), there is no indication of a systematic chloride concentration variation with elevation upon the individual component types. It is concluded from this that spatial property differences with location upon the individual members were responsible for the variation in chloride concentration with position. If this is the case, then it can be reasoned that at least localized areas of high permeability are likely to be present on any given member and that initial cracking and spalling should occur here.

Concrete Cover over ECR. An additional factor that influences the initiation time for corrosion induced concrete deterioration is cover over the embedded steel (see Equations 2 and 3). To investigate this, cover measurements were made upon the structure for both the reinforcement cage and bearing pad grid and according to the type of component (footer/strut, pile cap, or leg).

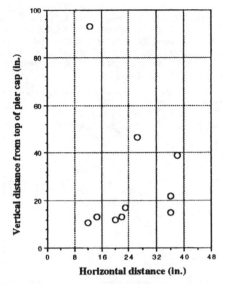

Figure 15: Core extraction locations in leg component.

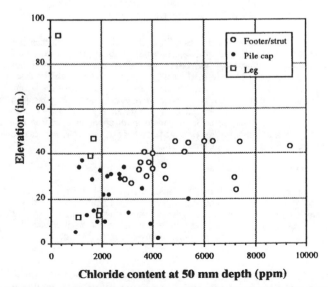

Figure 16: Chloride concentration at 50 mm depth along the elevation of each structural component.

Figure 17 presents the results of this as a cumulative cover distribution plot and shows this parameter to be distinct for each of the three components. Of particular interest is that, while the distribution for the leg and footer components was equally narrow, the former is displaced toward lower cover compared to the latter (mean leg cover 85 mm compared to 104 mm for the footers). On the other hand, cover for the pile caps was more distributed with a number of relatively low values being recorded. However, the mean cover for the pile caps was essentially the same as for the legs.

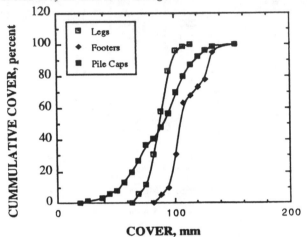

Figure 17: Plot of cummulative cover distribution
for different substructure components.

Figure 18 compares the cover distribution for the pile cap reinforcement cage reinforcement and for the bearing pad grid reinforcement in both the pile caps or footers and shows that, while the incidents of low cover were about the same in each case, the overall trend was such that the bearing grid steel was more shallow than for that of the reinforcement cage (mean cover was 69 mm for the former and 91 mm for the latter). The fact that concrete cracking and spalling was particularly evident in the bearing grid case, as noted above, was probably a consequence of this relatively low cover. Also, the observation that concrete cracking and spalling were more apparent upon the pile caps than for the footers and legs probably reflects, at least in part, the relatively low cover at the distribution extreme for the pile caps (see Figure 17).

Application of Data to the Damage Model

General. Based upon the results from the field inspection and from the laboratory analyses, sufficient information was considered available for application of the damage model, as described above and illustrated schematically in conjunction with Figure 1. Once this is accomplished, then sufficient information exists from which informed decisions could be made regarding substructure repair-rehabilitation-replacement strategies.

Period I (Time to Corrosion Initiation). Application of Equation 3 to determine TTC, as described above, requires statistical information regarding concrete cover and the chloride diffusion coefficient and knowledge of the chloride corrosion threshold. The procedure by which each of these was evaluated is described below.

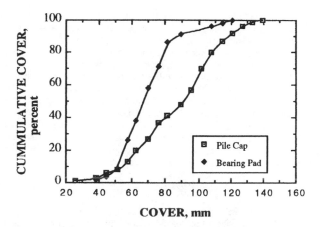

Figure 18: Comparison between cover distributions for
reinforcement cage and bearing pad grid.

Diffusion constant determinations.   From the chloride concentration-depth
information in Figures 9-11, diffusion constants were calculated using Equation 2 and
chloride concentrations which corresponded to the 97.5, 90, 80, 70, and 50 percent
probabilities of occurrence and for each of the three component types. Table 1 provides
the results from this and reveals that the footers were most permeable of the three to
chlorides and the legs were the least.

	COVER, mm.			
	Footer Reinf. Cage	Pile Cap Reinf. Cage	Leg Reinf. Cage	Bearing Pad Grid
Lowest 2.5%	84	38	66	32
Lowest 10%	95	51	74	51
Lowest 20%	99	64	77	53
Lowest 30%	102	71	81	58
Mean (50%)	105	89	86	66

Table 2: Concrete cover over ECR for the different components according to selected
occurrence probabilities.

Cover determinations for ECR. From the data in Figures 17 and 18 the concrete
cover over the ECR was determined for each of the three component types at probabilities
of occurrence of 2.5, 10, 20, 30, and 50 percent. These values are listed in Table 2.

	D (Footer), cm^2/sec	D (Pile Cap), cm^2/sec	D (Leg), cm^2/sec
Upper 97.5%	3.10E-07	1.43E-07	8.40E-08
Upper 90%	2.56E-07	1.21E-07	7.60E-08
Upper 80%	2.41E-07	1.14E-07	7.40E-08
Upper 70%	2.29E-07	1.11E-07	7.20E-08
Upper 50%	2.19E-07	9.80E-08	6.80E-08

Table 1: Diffusion constant for the different components according to selected
occurrence probabilities.

Chloride threshold determination. Several studies have indicated that the chloride concentration threshold for initiation of corrosion of bare reinforcing steel in concrete is in the 250-325 ppm range (Spellman and Stratfull, 1963; Clear, 1976); however, prior studies upon Florida structures have yielded values in the range 500-1500 ppm (Kessler and Powers, 1997). No previous studies have identified a chloride threshold for ECR. In the present study the requirement to determine the threshold was addressed in two ways: first, by considering that this value lies above the chloride concentration at the depth of uncorroded ECR samples that were acquired and analyzed as a part of the bridge coring and, second, by identifying sites of recent cracking/spalling on the various bridge components and analyzing for chloride concentration as a function of depth into the steel near these locations. From the diffusion constant and considering that the corrosion which gave rise to any recently disclosed incidents of concrete surface cracking initiated 3.5 years earlier (the beginning of Period 2 in Figure 1 or at a structure age of 12.5 years, as explained above) the chloride concentration at the steel depth at that time, which is assumed to be the threshold value, was calculated from Equation 2.

Figure 19 plots chloride concentration at the ECR depth versus cover* for the cores which contained ECR and identifies the ECR for which corrosion was apparent. It is concluded from these data that the chloride threshold for the footers was between 2,375 and 5,188 ppm, while for the pile caps it is greater than 1,469 ppm. Corrosion induced spalling associated with the footer reinforcement cage was rare; however, a pair of fine, hairline cracks were disclosed upon the north footer of Bent 76. A core from this general location revealed the ECR to be corroded, and the chloride analysis results constitute the extreme data point in Figure 19.

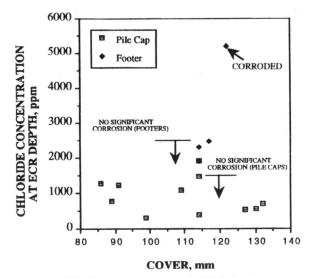

Figure 19: Chloride concentration at the ECR depth versus cover
for core samples which contained reinforcement.

---

* No significance is attached to the cover parameter being included as an axis representation in this figure, but it simply provides an added point of information.

Also, the most elevated chloride profile in Figure 6 is for this same core. The diffusion constant calculated from this using Equation 2 and assuming an exposure time of 16 years was $6.14 \cdot 10^{-7}$ cm^2/sec. Note that this value is well above the upper 97.5 percent probability of occurrence value of $3.10 \cdot 10^{-7}$ cm^2/sec, as listed in Table 1. From the calculated diffusion coefficient (D = $6.14 \cdot 10^{-7}$ cm^2/sec), Equation 2 was employed again; and a chloride concentration of 4,900 ppm was calculated for the ECR depth (122 mm) at year 12.5 (the projected time of corrosion initiation, see above); and this is taken as the threshold for the footer components. Such a value is higher than expected, but it is consistent with the fact that there were relatively few instances of corrosion induced cracking for the footer ECR reinforcement cage despite high levels of chlorides. Possible explanations are that 1) these components are water saturated such that little dissolved oxygen is available at cathodic sites and 2) the ECR is electrically connected to the drilled shaft steel and that the latter is providing cathodic protection to the former.

Because of the abnormally high chlorides in the footer of Bent 76, a petrographic analysis was performed upon this core. This revealed the concrete constituents to be generally in accord with the mix design specification, and the microstructure was comparable with concrete samples from others locations on the bridge. It was concluded that sea water may have been employed in the mixing of this particular footer concrete, thereby elevating the chlorides accordingly.

The finding that chlorides at a particular depth in the pile caps and legs were below those of the footers (see Figures 6-11) is consistent with the former two component types being precast and the latter cast-in-place and with the fact that the footers are more directly exposed to sea water. To identify the chloride threshold for the pile caps, a location of recent cracking was disclosed at Bent 5; and based upon the chloride profile of a core acquired from this vicinity, a diffusion constant of $6.14 \cdot 10^{-8}$ cm^2/sec was calculated. Assuming that the corrosion which led to this crack began 3.5 years earlier, as discussed above, a chloride concentration at the steel depth (70 mm) of 1,700 ppm was determined; and it is assumed that this corresponds to the threshold. Note that this value is consistent with the data in Figure 19 in that it is slightly above the highest chloride concentration measured at the depth of uncorroded ECR from pile cap cores. This same threshold, 1,700 ppm, is assumed to apply also to the legs.

Analysis. The results from the probability of occurrence determinations for the diffusion constant and cover (Tables 1 and 2, respectively) were combined in all possible combinations to establish a probability matrix that a particular cover would occur in conjunction with a specific D. The time-to-corrosion was then calculated for each of these combinations using the chloride threshold as determined for the component in question (see above) and the average surface chloride concentration for the component type. Tables 3-7 present the results of this for 1) the footer reinforcing cage ECR, 2) either the footer or pile cap bearing grid ECR assuming the D values for the pile cap concrete (see Table 1), 3) the footer bearing grid ECR assuming the D values for the footer concrete, 4) the pile cap reinforcement cage ECR, and 5) the leg reinforcement cage ECR, respectively. In these tables the first number in a particular cell is the calculated TTC in years (extent of Period 1 of Figure 1), and the second is the number of piers upon which the particular D and cover combination should be realized. For example, the first cell in Table 3 indicates that the time-to-corrosion for the footer reinforcement grid ECR with D = $3.10 \cdot 10^{-7}$ cm^2/sec (upper 97.5% case) and a cover of 84 mm (the lowest 2.5% case) is 25 years; however, the number of piers for which this D and cover combination is projected to be less than one, meaning that this situation is not likely to occur. On the other hand, a combination of this same D and a cover of 99 mm (20% cover case) should be realized on one footer. The calculation was performed in the case of the footer bearing pad grid using both the footer concrete and pile cap diffusion constants, and appropriateness of one versus the other is discussed subsequently.

D, cm^2/sec	COVER, mm				
	84 (2.5%)	95 (10%)	99 (20%)	102 (30%)	105 (mean)
3.10E-7 (upper 97.5%)	25//<1	33//<1	36//1	37//1	40//1
2.56E-7 (upper 90%)	51//<1	66//1	72//2	75//3	81/5
2.42E-7 (upper 80%)	77//<1	96//2	108//3	113//5	122//8
2.29E-7 (upper 70%)	108//<1	140//2	-	-	-
2.19E-7 (mean)	214//1	-	-	-	-

First number in cell is time-to-corrosion in years. Second number is the number of piers of the Long Key Bridge upon that particular combination will be realized.

Table 3: Time-to-corrosion for the footer reinforcement cage ECR for various diffusion constant and ECR cover combinations.

D, cm^2/sec	COVER, mm				
	32 (2.5%)	51 (10%)	53 (20%)	58 (30%)	66 (mean)
1.43E-7 (upper 97.5%)	1.3//<1	3.3//<1	3.6//1	4.3//1	6//1
1.21E-7 (upper 90%)	1.8//<1	4.6//1	5//2	6//3	8//5
1.14E-7 (upper 80%)	2.0//<1	5//2	6//3	7//5	9//8
1.11E-7 (upper 70%)	2.2//<1	6//2	6//4	8//5	9//8
9.80E-8 (mean)	2.6//<1	7//2	7//5	9//6	11//9

First number in cell is time-to-corrosion in years. Second number is the number of piers of the Long Key Bridge upon which a particular combination will be realized.

Table 4: Time-to-corrosion for various diffusion constant and ECR cover combinations in the case of footer or pile cap bearing pad ECR and using the pile cap concrete D and chloride threshold values.

D, cm^2/sec	COVER, mm				
	32 (2.5%)	51 (10%)	53 (20%)	58 (30%)	66 (mean)
3.10E-7 (upper 97.5%)	4//<1	9//<1	10//1	12//1	16//1
2.56E-7 (upper 90%)	7//<1	19//1	21//2	25//3	32//5
2.42E-7 (upper 80%)	11//<1	28//2	31//3	37//5	48//8
2.29E-7 (upper 70%)	15//<1	40//2	44//4	52//5	67//8
2.19E-7 (mean)	30//<1	78//2	87//5	104//6	133//9

First number in cell is time-to-corrosion in years. Second number is the number of footers of the Long Key Bridge upon that particular combination will be realized.

Table 5: Time to corrosion for various diffusion constant and ECR cover combinations in the case of footer bearing pad ECR and using the footer concrete D and chloride threshold values.

Figure 20 presents the TTC results graphically as a plot of the number of piers upon which corrosion is projected to initiate as a function of time. Thus, the model predicts the first occurrence of corrosion as having transpired upon both the bearing grid and the pile cap reinforcement cage after four and five years, respectively, with the rate of subsequent occurrences being slightly greater with time for the former, although the difference is probably within the margin of error considering the assumptions that were made. If the footer diffusion constant and chloride threshold apply to the reinforcement grid in these components, then corrosion initiation is projected to occur

D, cm^2/sec	COVER, mm				
	38 (2.5%)	51 (10%)	64 (20%)	71 (30%)	89 (mean)
1.43E-7 (upper 97.5%)	1.8//<1	3.3//<1	5//1	6.4//1	10//1
1.21E-7 (upper 90%)	2.6// <1	4.6//1	7//2	9//3	14//5
1.14E-7 (upper 80%)	2.9//<1	5//2	8//3	10//5	16//8
1.11E-7 (upper 70%)	3.1//<1	6//2	9//4	11//5	17//8
9.8E-8 (mean)	3.8//<1	7//2	10//5	13//6	21//9

First number in cell is time-to-corrosion in years. Second number is the number of
pile caps of the Long Key Bridge upon which a particular combination will be realized.

Table 6: Time to corrosion for various diffusion constant and ECR cover combinations
in the case of the pile cap reinforcement cage ECR.

D, cm^2/sec	COVER, mm				
	66 (2.5%)	74 (10%)	77 (20%)	81 (30%)	86 (mean)
8.4E-8 (upper 97.5%)	19//<1	24//<1	26//1	29//1	32//1
7.60E-8 (upper 90%)	25//<1	31//1	35//2	38//3	43//5
7.40E-8 (upper 80%)	29//<1	36//2	40//3	44//5	49//8
7.20E-8 (upper 70%)	32//<1	40//2	44//4	49//5	55//8
6.80E-8 (mean)	38//<1	48//2	53//5	58//6	66//9

First number in cell is time-to-corrosion in years. Second number is the number of
legs of the Long Key Bridge upon which a particular combination will be realized.

Table 7: Time to corrosion for various diffusion constant and ECR cover combinations
in the case of the substructure legs.

at year ten; and the rate at which further instances occur is more modest than if the pile
cap diffusion constant and threshold apply.   These incidents are followed by corrosion
initiation of ECR in the leg components at year 29 and upon the footer reinforcement cage
at year 36. Figure 21 provides a replot of the short time pile cap and bearing grid data
from Figure 20 on an expanded time scale.

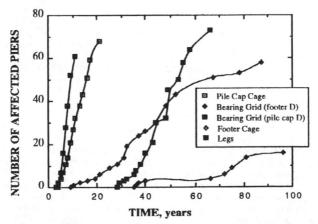

Figure 20: Projected time-to-corrosion as a function of time for
various substructure components.

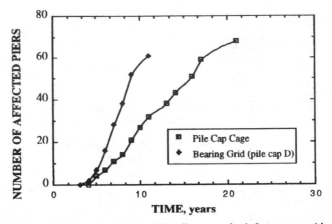

Figure 21: Expanded view of the pile cap and reinforcement grid
data from Figure 18.

Periods II and III (Time from Corrosion Initiation to Limit State).

As already noted, the time once corrosion has initiated for a crack to appear at the
concrete surface was assumed to be 3.5 years. Determination of Period III is more
complex, however, and involves definition of the limit state. In the case of the Long Key
Bridge the individual piers are separate structural members, since each acts independently
from the others. Also, the limit state for the individual components comprising a particular
pier (footer/strut, pile cap and legs) must be addressed individually, since failure of any
one of these compromises a particular pier and, in turn, the overall bridge.

Based upon the damage forms (cracking and spalling) that have developed to-date,
three potential failure modes were identified that can result in a limit state being achieved,
as listed below:

1. Cracking and spalling in the central portion of the pile caps to an extent that the
   middle pair of the four horizontal, longitudinal cage ECRs is affected.
2. Edge cracking and spalling of either the pile cap or footer (or both) to an extent
   that this undercuts the bearing pad area.
3. Cracking and delamination along the reinforcement grid either above or below the
   bearing pad and corrosion of this steel to an extent that one-half of the metal cross
   section is consumed.

Figure 22 provides a schematic illustration of the first two. In these cases it is concrete
section loss and the resultant reduced load bearing capacity that is determinant. In
instances of delamination along the reinforcement grid (limit state 3) the load bearing
capacity of the concrete per se may be retained, but corrosion of the ECR where this is
exposed via cracks is considered critical. In addition, cracks could develop from the grid
steel which extend into the primary section of the concrete. Other limit states are also
possible (for example, cracking and spalling of legs), but every indication to-date is that
these will be preempted by one or more of the three modes cited above. This particular
point is confirmed by the failure model, as explained subsequently. If, however, successful
mitigation techniques are applied which forestall failure according to the three above
modes, then one or more alternative modes may become important.

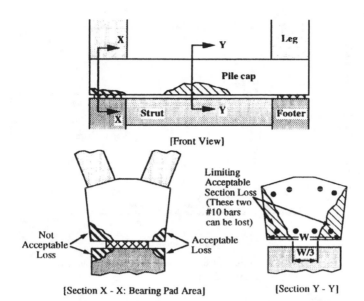

Figure 22: Schematic illustration of various limit states (failure conditions) for pile cap and footer components.

The time of Period III was evaluated by considering that the first incident of cracking was discovered in 1986 in association with the reinforcement cage of the Pier 6 pile cap (Kessler and Powers, 1987) and that this same component is now approaching its limit state some 11 years later. On this basis, a time of 12 years for Period 3 was assumed. In the case of limit state three, however, Period III corresponds to the time for one-half of the bar cross section to be consumed, as noted above. Based upon the fact that the grids are comprised of 9.5 mm bars and assuming a corrosion rate at the end of Period II of 0.13 mmpy, the time for Period III is 11 years. Thus, each of the three limit states considered require about the same time.

The net time to achieve a limit state is then TTC + 3.5 yrs + 12 yrs, where data for the first term (TTC) are obtained from Tables 3-7. The results are plotted in Figure 23. In both cases results for the footer bearing pad grid based upon the footer diffusion constant and chloride threshold have been omitted, since the field observations regarding how damage has actually progressed with time are more in agreement with the pile cap diffusion constant and threshold trend. A possible explanation for why a different diffusion constant apparently applies to the footer grid compared to the cage is that the concrete may be dryer in the vicinity of the former (grid) because of its proximity to the top surface. Consequently, the moisture content of the concrete here is more like that of the pile cap than for the lower elevations of the footer. Based upon the results in Figure 23, a projection has been made regarding the rate at which failure of the different substructure components should progress. Figure 24 replots the data in Figure 23 with the failure progression being represented by a linear trend with the corresponding equation for

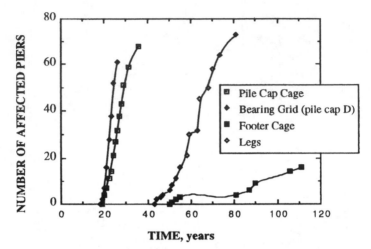

Figure 23: Time-to-failure of the various substructure components as a function of time.

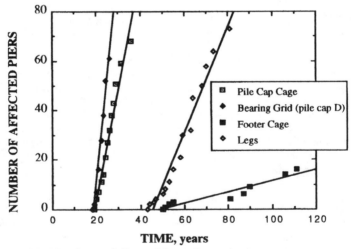

Figure 24: Net time-to-failure of the various substructure components assuming that the number of failures increases linearly with time.

Information of the type presented in Figures 23 and 24 and Table 9 provides a quantification of the time and rate of failure of the various Bridge components. This, in turn, gives bridge engineers the input information that is necessary in order for informed decisions to be rendered regarding repair, rehabilitation, and partial and full replacement.

COMPONENT TYPE	GOVERNING EQUATION	R^2
Bearing Grid	NAP = -159.4 + 8.55TTF	0.97
Pile Cap Cage	NAP = -85.86 + 4.51TTF	0.97
Legs	NAP = -97.57 + 2.15TTF	0.967
Footer Cage	NAP = 10.86 + 0.223TTF	0.899

NAP = Number of affected piers.
TTF = Time to failure in years.

Table 8: Damage rate equations for each of the component types based upon the trend lines in Figure 24.

COMPONENT TYPE	YEARS TO FIRST FAILURE	YEAR OF FIRST FAILURE*	NUMBER OF PIER FAILURES PER SUBSEQUENT YEAR
Bearing Grid	19	1999	9
Pile Cap Cage	19	1999	5
Legs	45	2025	2
Footer Cage	49	2029	0.2

* Assumes exposure began in 1980.

Table 9: Projected time and year of initial failure and the rate at which subsequent failures occur for each component.

Based, in part, upon the present approach and analysis a decision was made to apply an arc sprayed zinc sacrificial anode cathodic protection (Kessler et al., 1990; Powers et al., 1992, Sagues and Powers, 1994) to the affected areas* with the expectation that such a system will provide at least partial protection for seven years. It is projection of 12 years may have been affected by this. Alternately, some of these components may have been cracked to such an extent that this cathodic protection has had little or no affect anticipated that at least two and perhaps three successive applications can be made prior to a limit state being reached.

Bibliography

Bamforth, P. B., 1993, "Concrete Classifications for R. C. Structures Exposed to Marine and Other Salt-Laden Environments," Proc. Fifth Int'l. Conf. on Structural Faults and Repair, Vol. 2, p. 31.

Cady, P. D. and Gannon, E. J., 1992, "Condition Evaluation of Concrete Bridges Relative to Reinforcement Corrosion: A Methods Application Manual," Report No. SHRP-S-330, Strategic Highway Research Program, National Research Council, Washington, DC.

---

* Sacrificial arc sprayed zinc cathodic protection has been in place on some of the Long Key Bridge substructure components for approximately five years, and the Period III time.

Clear, K. C., 1976, "Time-to-Corrosion of Reinforcing Steel in Concrete Slabs," FHWA-RD-76-70, Federal Highway Administration, Washington, DC.

Kessler, R. J. and Powers, R. G., 1986, "Corrosion Evaluation of Substructure Cracks in Long Key Bridge," Report No. 86-3, Corrosion Research Laboratory, Bureau of Materials and Research, Florida Department of Transportation, Gainesville, FL.

Kessler, R. J. and Powers, R. G., 1987, "Corrosion Evaluation of Substructure Long Key Bridge," Report No. 87-9A, Corrosion Research Laboratory, Bureau of Materials and Research, Florida Department of Transportation, Gainesville, FL.

Kessler, R. J., Powers, R. G., and Laska, I. R., 1990, "Zinc Metallizing for Galvanic Cathodic Protection of Steel Reinforced Concrete in a Marine Environment," paper no. 324 presented at CORROSION/90, Las Vegas.

Kessler, R. J. and Powers, R. G., 1997, Corrosion Research Laboratory, Bureau of Materials and Research, Florida Department of Transportation, Gainesville, FL, unpublished research.

Powers, R. G., Sagues, A. A., and Murase, T., 1992, "Sprayed-Zinc Galvanic Anodes for the Cathodic Protection of Reinforcing Steel in Concrete," Proc. Materials Engineering Congress,, Ed. T. D. White, Am. Soc. Civil Engrs., New York, p. 732.

Purvis, R. L., Babaei, K., Clear, K. C., and Markow, M. J., 1994, "Life-Cycle Cost Analysis for Protection and Rehabilitation of Concrete Bridges Relative to Reinforcement Corrosion," Report No. SHRP-S-377, Strategic Highway Research Program, National Research Council, Washington, DC.

Sagues, A. A., Powers, R. G. and Kessler, R. J., 1995, "An Up-Date on Corrosion Precesses and Field Performance of Epoxy-Coated Reinforcing Steel," Proc. COST-509 Workshop on Corrosion and Protection of Metals in Contact with Concrete, European Co-operation in the Field of Scientific and Technical Research, Seville, Spain.

Sagues, A. A. and Powers, R. G., 1994, "Low-Cost Sprayed Zinc Galvanic Anode For Control of Corrosion of Reinforcing Steel in Marine Bridge Substructures," Final Report submitted to Strategic Highway Research Program on Contract No. SHRP-88-ID024 by Univ. South Florida.

Spellman, D. S. and Stratfull, R. F., 1963, "Concrete Variables and Corrosion Testing," Highway Research Record, No. 423, p. 16.

Weyers, R. E., Prowell, B. D., Sprinkel, and M. M., Voster, 1993, "Concrete Bridge Protection, Repair, and Rehabilitation Relative to Reinforcement Corrosion: A Methods Application Manual," Report No. SHRP-S-360, Strategic Highway Research Program, National Research Council, Washington, DC.

Corrosion Service Life Model Concrete Structures

Richard E. Weyers[1]

Abstract

Bridges are a major element in ground transportation networks. The magnitude of the backlog of bridge rehabilitation needs is well known. A major proportion of the backlog work is attributed to the corrosion of steel in concrete. The chloride corrosion of steel in concrete structures is not confined to bridges but also reduces the service life of parking garages and sea and coastal structures.

The service life model for reinforced concrete structures in chloride laden environments consists of the following serial phases: diffusion to the depth of the steel that would precipitate first maintenance actions; corrosion of the steel at the first maintenance depth until cracking and spalling occurs; continue spalling until a damage level is reached which is defined as the end-of-functional service life for an element or structure. The diffusion phase is described by Fick's Law and the boundary conditions define the solution form. The time-to-cracking model is dependent upon factors as concrete strength properties, cover depth and corrosion rate. Corrosion rate has a significant influence on the time-to-cracking which typically occurs in three to seven years after initiation.

The utility of the model in estimating the service life of corrosion protection systems as increased cover depth, corrosion inhibitors and low permeable concrete is presented. Also the means for determining the time-to-first maintenance and annual maintenance requirements to the end-of-functional service life are presented. The model in conjunction with a life-cycle cost model will enable the selection of the least cost solution to corrosion protection.

---

[1] Professor of Civil Engineering and Director of the Center for Infrastructure Assessment and Management, 200 Patton Hall, Virginia Tech, Blacksburg, VA 24061-0105.

Introduction

Bridges are a major element in ground transportation networks. Of the 574,671 highway bridges over 6 meters in length in the United States, 32% are structurally deficient or functionally obsolete. The average roadway separation distances between bridges is 6.4 km. Whereas, the average bridge detour is 32 km. Often lost revenues in delay times, increased transportation costs, reduced productivity and impact on local businesses exceed the replacement cost of a bridge within one year.

A major portion of the backlog bridge work in the United Sates is attributed to the corrosion of steel in concrete. Since the identification of the cause of the corrosion deterioration of concrete bridges in the early 1960's, numerous methods have been developed to extend the service of bridges in chloride laden environments. Given the magnitude of the current backlog of bridge work, the selection of maintenance, repair, rehabilitation, and replacement methods requires the solution to be cost effective. Cost effectiveness by definition is minimum life-cycle cost. In addition to the price of alternative solutions and interest rates, life-cycle cost analysis requires the service life of alternative solutions.

This paper presents a service life model for the corrosion of bridge components in chloride laden environments. Although the model was developed for bridge decks, the methodology is applicable to all concrete structures in chloride laden environments as coastal and sea structures and parking garages.

Service Life Model

The cumulative damage curve for structures is an ogive, or S-shaped curve. Weyers, and et. al. (1993) identified the 4 regions of the cumulative corrosion deterioration curve for concrete bridges as the initial ill-defined rate, initial and final near-steady rate and the reducing rate. For concrete bridge components, the initial and final near-steady-state deterioration rates are fairly predictable because (a) the rate of chloride diffusion depends on the concrete's permeability and ambient temperature; (b) the amount of chloride exposure is a function of the environmental conditions as the mean average snowfall, brackish water chloride content, seawater or distance from the coast; (c) construction methods result in a predictable (normal) distribution of the concrete cover depths; (d) corrosion rate is dependent on given environmental exposure conditions. Cady and Weyers (1984) presented the service model for bridge decks as being composed of four serial time periods (1) rapid initial damage as the result of construction faults (subsidence cracking); (2) chloride diffusion period for initiation of corrosion at the lowest 2.5 percentile concrete cover; (3) corrosion of the steel at the 2.5 percentile cover depth until cracking of the cover concrete occurs; (4) a uniform deterioration rate until the damage level defined as the end-of-functional service life is reached, see Figure 1. The model allowed for the prediction of the time to first maintenance and the level of maintenance required

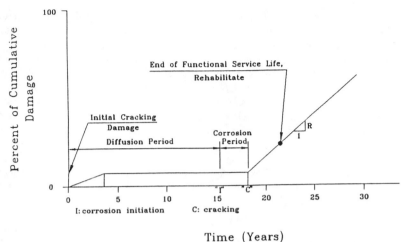

Figure 1. Chloride Corrosion Deterioration Process for a Concrete Element

to the time for rehabilitation. For the comparison of alternative service lives, only the time to initial corrosion at the depth of steel defined as end-of-functional service life and the corrosion time for cracking at this depth are required.

The following presents the development of the model for concrete bridge decks constructed with bare steel. The application of the model shall then be extended to the use of low permeable concretes, corrosion inhibitors and surface sealers and coatings. In the development of service life models, the first measurable parameter that needs to be defined is the end point or the end-of-functional service life.

End-of-Function Service Life

In the development of a definition for the end-of-functional service life the governing condition must first be identified. Once the governing condition is determined a set of measurable parameters must be selected and the magnitude of the parameters selected as the end-of-functional service life. The condition which defines the end-of-functional service life may be a reduction in structural capacity or an other parameter as riding quality or safety as in the case of bridge decks. Thus end-of-functional service is component and structure dependent. As an example, a reduction in structural capacity of 20 percent may be acceptable for a given structural element because it is still able to provide an acceptable level of service. Whereas, for another element a 5 percent reduction in capacity may define the end-of-functional service life. In either case, the governing condition is structural capacity or safety and the magnitude may be calculated or otherwise determined.

For bridge decks riding safety governs, because long before the structural safety of the deck is of concern, riding quality has deteriorated to the point of reducing traffic speeds and driving safety through possible loss of vehicle control. Riding quality is a more subjective parameter than structural integrity. Thus, engineering opinion was used to define the end-of-functional service life. Fitch (1995) developed a set of visual deteriorated conditions of bridge decks and survey questionnaires to define the end-of-function service life for bridge decks. The survey kits were sent to 90 qualified engineers and 60 responded. Analysis of the responses demonstrated that the worst traffic lane defined the condition of end-of-functional service life. The measurable parameter was the sum total for spalls, delaminations and asphalt patches and the magnitude between 9.3 to 13.6 percent of the worst deteriorated traffic lane. The average is approximately 12 percent of the worst deteriorated lane which is approximately the 12 percentile least concrete cover depth. For bridge decks, the reinforcing steel cover is normally distributed and the standard deviation is 9.5 mm.

Time-to-Initiate Corrosion

The penetration of chloride ions into porous concrete may be described by Fick's Law

$$\partial C/\partial_t = D_c(\partial c^2/\partial x^2) \tag{1}$$

where

C = chloride concentration
x = depth
t = time and
D_c = diffusion constant.

The diffusion constant, D_c, is a function of the temperature, porosity, pore size distribution and continuity of the pore structure. For concrete, the capillary pore size distribution and the aggregate - cement paste transition zone are of interest. The size and distribution of these pores are a function of the water to cement ratio, type and amount of pozzolan and the consolidation of the concrete. Ambient temperatures are structure location dependent. For example, in the northern states the capillary pore water may be frozen for two to three months of the year. Whereas in the southern states the capillary pore water will be liquid throughout the year and the temperatures are significantly higher during the summer months. Thus, diffusion constants determined from field structures are referred to as effective diffusion constants.

The solution to Fick's Law is dependent on the surface boundary conditions. Weyers, and et. al. (1994) showed that the increase in surface chloride approached the square root of time function.

$$C_{(o,t)} = S\sqrt{t} \tag{2}$$

where

$C(o,t)$	=	the near surface chloride concentration
S	=	a surface chloride concentration coefficient
t	=	time

The concrete chloride surface increase being a function of the square root of time may be more applicable to constant chloride concentration exposure conditions for marine structures in brackish or seawater than for bridge decks. For bridge decks where chloride exposure is during the winter months and rainwater during the spring, summer, and fall months, the underlying phenomenon of near surface chloride increase is generalized in Figure 2. Since the increase in chloride in the near surface is relatively rapid. A near constant chloride content is reached in about five years, compared to the service life of a bridge decks of 20 to 40 years, an approximation of a near constant chloride at a depth of 13 mm is reasonable. The solution to Fick's Law for a constant chloride boundary condition is as follows:

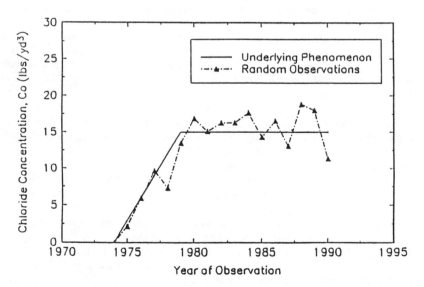

(Note: 1 lbs/yd³ = 0.5933 kg/m³)

Figure 2. Generalized Concept of a Constant Near-Surface Chloride Concentration

$$C_{(x,t)} = C_o \left[1 - \text{erf} \frac{x}{2\sqrt{D_c t}} \right] \tag{3}$$

where

$C_{(x,t)}$ =   the chloride content at depth x and time t
C_o =   near surface constant chloride content.

However, Zemajtis (1996) showed that the increase in surface chloride content is a function of the square root of time for concrete treated with surface sealers and coatings. The square root of time function, $S\sqrt{t}$, is a measure of the efficiency of the sealer and coating in excluding chlorides.

The near surface chloride content, C_o in Equation 3, is determined from measurements on structures. For structural elements continuously exposed to chloride ions, the surface chloride may be estimated from measurements of samples from 0 to 6 mm or 0 to 13 mm and the average chloride content for the depth range projected back to the surface. However, the chloride penetration of the surface concrete, 0 to 13 mm, may not be strictly diffusion, but may be highly influenced by surface cracking as drying shrinkage cracking and/or capillary suction where cyclic wetting and drying conditions exist.

For bridge decks, the chloride content is relatively constant at a depth of 13 mm and the near surface chloride contents are measured for samples taken from 6 to 20 mm. Weyers, and et. al. (1994) showed that the near surface chloride contents were related to the environmental exposure conditions in the United States. Corrosion environments were categorized as low, moderate, high and severe with corresponding surface chloride content ranges of 0 to 2.4; 2.4 to 4.7; 4.7 to 5.9; and 5.9 to 8.9 kg/m^3.

The effective chloride diffusion constant for a structural element is determined from Equation 3. Chloride contents for a number of locations as a function of depth are measured. The sum of errors squared are determined for estimated effective diffusion constant values. The minimum sum of errors squared is the best fit curve for a location. However, the diffusion constant for a structural element is not merely the average of the best fit diffusion constants. Some locations do not show a clear minimum sum of errors squared. Also, a simple average will result in giving too much weight to a single high value. Thus, the effective diffusion constant for a structural element as a bridge deck is the minimum of the sum of sum of errors squared. Figure 3 illustrate the above procedure for determining the effective diffusion constant. Weyers, and et. al. (1993) present guidelines for the number and sample depths at a location and number of locations for structural components.

(Note: 1 in²/yr = 6.4516 cm²/yr)

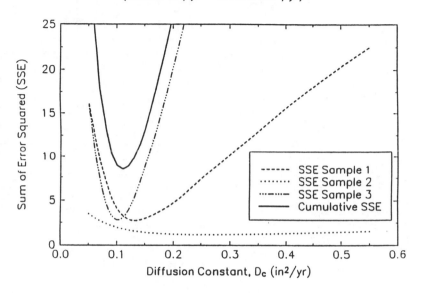

Figure 3. Least Squared Fit Concept

For the analyses the chloride content at depth x and time t, $C_{(x,t)}$, in Equation 3 is set equal to the chloride corrosion threshold limit. The most used value is 0.71 kg/m³ but varies depending upon various conditions. A discussion of the chloride corrosion threshold limit is outside of the scope of this paper but one point is worthy of mentioning. It is well known that corrosion initiation of materials that are protected by passive layers as steel in concrete is normally delayed. Thus, there is some time required to destroy the passive layer at the threshold concentration. To adjust for this factor the depth, x, is adjusted to 13 mm below the top of the reinforcing steel. The adjustment also simplifies the corrosion initiation calculation because the depth, x, becomes the concrete cover depth.

Weyers and et. al. (1993) developed surface chloride concentrations and effective chloride diffusion constants from data submitted by a number of State Department of Transportation, see Table 1. For the states in the northern climate, the surface chloride concentrations reflect both the winter maintenance policy of the state and the amount of annual snowfall. For example, the annual snowfall for New York is significantly greater than Kansas and New York's rate of deicer salt usage in kg/lane-km/yr is about twice that of Kansas. The surface chloride content of New York is about four times that of Kansas, see Table 1. However, the effective diffusion constants for Kansas and New York are approximately equal indicating that

Table 1. Surface Chloride Concentration, C_o, and Effective Chloride
Diffusion Constants for Various States

State	Mean C_o kg/m^3	Mean D_c mm^2/yr
Delaware	5.2	32
Minnesota	3.9	32
Iowa	4.8	32
West Virginia	5.1	45
Indiana	5.4	58
Wisconsin	6.1	71
Kansas	2.2	77
New York	8.8	84
California	1.9	161
Florida	3.6	213

the influence of annual temperature and material properties are about equal. As shown in Table 1, Florida and California have the highest chloride diffusion constant as would be expected for the warmer climates of these states. Florida's surface chloride content represent brackish and seawater conditions, whereas California's surface chloride content represent both coastal and high mountain areas.

Time-to-Cracking

Upon initiation of corrosion, the increase in volume of the corrosion products creates sufficient internal pressure to crack the cover concrete. According to Faraday's Law, the metal weight loss is directly proportional to the corrosion current.

$$\text{Weight of metal reacting} = kIt \qquad (4)$$

where

I = current in amperes
t = time in seconds
k = electrochemical equivalent constant for a given metal reaction in grams/ampere/second

For ion Fe going to Fe^{++}, k is equal to 2.89×10^{-4} g/amp-sec.

However, the iron ion does not immediately react to produce an iron corrosion product. The iron must diffuse through the rust layer and then react to produce an oxidized form of iron. The corrosion products have different densities and thus their expansion create different internal pressures. Also, not all corrosion products create pressure but some expand in the pore space in the transition zone between the steel bar and cement paste, entrapped air voids and water pockets, or diffuse through the capillary pores and oxidize in porous areas without creating pressures.

There are three stages to the time-to-cracking model which are as follows:

1. Free expansion. The passive film is destroyed by the chloride ions, the metallic iron at the anode is oxidized to form ferrous ions. The ferrous ions react to produce ferrous hydroxide which then can be converted to hydrated ferric oxide. The reaction products expand in the free space. The volume of the free space is directly related to the surface area of the reinforcement, w/c+p ratio, degree of hydration and degree of consolidation. The total amount of corrosion products, W_T, accumulate until the free space is filled with corrosion products, W_p.

2. As the total amount of corrosion products, W_T, exceed W_p, internal pressure is created. The internal pressure increases as W_T increases beyond W_p.

3. When the internal pressure exceeds the tensile strength of the concrete, W_t is equal to the critical amount of corrosion products, W_{crit}. The critical weight of corrosion products, W_{crit}, required to crack the cover concrete is related to the tensile strength of the concrete and the cover depth. Generally values of W_{crit} increase with increasing cover depth and concrete strength. However, this may not be the case for high strength concrete because of the reduction in pore space surrounding the steel and in the bulk cement paste.

Liu (1996) showed that the critical weight of corrosion products, W_{crit}, may be expressed as follows:

$$W_{crit} = \rho_{rust}\left[\pi(d_s + d_o)D + \frac{W_{st}}{\rho_{st}}\right]$$ (5)

where

ρ_{rust} = the density of the corrosion products

ρ_{st} = the density of steel

d_o = the thickness of pore band around the steel/concrete interface

d_s = the thickness of corrosion products needed to generate tensile stresses

D = the diameter of steel reinforcing

Liu (1996) also showed that the time-to-cracking is equal to the following expression:

$$t_{cr} = \frac{W_{crit}^2}{2k_p}$$ (6)

where

k_p = 0.106 $(1/\alpha)\pi D\ i_{corr}$
α = 0.57, the mean of the corrosion products $F_e(OH)_2$ and $F_e(OH)_3$
 where α is equal to 0.523 and 0.622, respectively.
D = diameter of the reinforcing steel, mm.
i_{corr} = annual mean corrosion current density, $\mu A/cm^2$

and

$$W_{crit}^2\, \rho_{rust}\left[\pi\left(\frac{Cf_t}{E_{ef}}\right)\left(\frac{a^2+b^2}{b^2-a^2}\right)+v_c+d_o\right]D+\frac{W_{st}}{\rho_{st}} \qquad (7)$$

where

ρ_{rust}	=	density of iron oxide, $3600 kg/m^3$.
C	=	cover depth
f_t	=	tensile strength of concrete
E_{ef}	=	$E_c/1+\psi_{cr}$
E_c	=	elastic modulus in compression
ψ_{cr}	=	creep coefficient
a	=	$D + 2d_o/2$
d_o	=	thickness of the pore band around the concrete/steel interface, 12.5μm
b	=	$C + D + 2d_o/2$
v_c	=	Poisson's ratio

For the following set of experimental parameters, f_t = 3.3 MPa, E_c = 27GPa, ψ_{cr} = 2.0, v_c = 0.18 and ρ_{rust} and d_o assumed to be 3600 kg/m^3 and 12.5 μm, respectively, Liu (1996) showed that experimental observed time to cracking was within the range of calculated time to cracking using α values of 0.523 ($F_e(OH)^2$) and 0.622 ($F_e(OH)^3$), see Table 2.

Table 2. Model Predicted Times to Cracking and Observed Times to Cracking

Test Series	Bar Diameter cm	Cover Depth cm	Measured Current Density $\mu A/cm^2$	Model t_r yr	Observed t_r yr
18512.0	1.6	2.72	3.50	0.56 - 0.75	0.72
2859.6	1.6	4.75	2.18	1.53 - 2.06	1.84
3859.6	1.6	6.95	1.67	3.34 - 4.49	3.54

Figure 4 presents the influence of the rate of corrosion on the time-to-cracking from the initiation of corrosion for the typical bridge deck parameters presented above. As

(Note: 1 in. = 2.54 cm &
1 mA/ft^2 = 1.08 µA/cm^2)

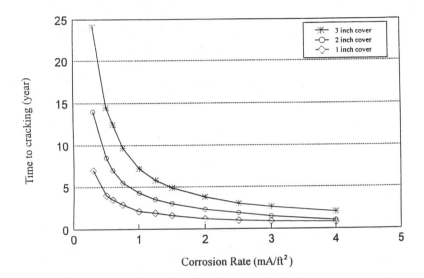

Figure 4. Effects of Corrosion Rate and Cover Depths on the Time to Cracking

shown, for annual corrosion rates greater than 1 µA/cm^2, the condition of almost all field corrosion measurements, the time-to-cracking is relatively small, less than 7 years for cover depths 76 mm and less. For annual corrosion rates greater than 2 µA/cm^2, the time-to-cracking is less than 5 years for cover depths 76 mm and less.

The annual corrosion rates presented in Figure 4 were calculated from weight loss measurements determined at the time of observed cracking. For the same series of test specimens, Liu (1996) measured the monthly corrosion current densities (corrosion rate) using two linear polarization devices, the unguarded 3LP and the Gecor device with a guard ring. The monthly measurements were integrated over time and the average annual corrosion rate calculated. Table 3 presents the average annual weight loss and device measurements. As shown in Table 3, the 3LP appears to over estimate the corrosion rate and the Gecor underestimate the corrosion rate. However, the measured corrosion rates are a snapshot of the monthly corrosion rate

Table 3.　Average Annual Corrosion Rates for Weight
Loss and Device Measurements.

Test Series	Exposure Period yrs	Average Annual Corrosion Rate $\mu A/cm^2$		
		Weight Loss	3LP	Gecor
18512.0	0.87	3.77	8.63	0.62
2859.6	1.84	2.34	8.19	0.50
3859.6	3.67	1.80	4.99	0.39

which varies with ambient temperature and other corrosion controlling factors. Liu (1996) demonstrated that for specimens outdoors in Blacksburg, VA the rate of corrosion may be expressed as follows:

$$\ln i = 8.45 + 0.77 \ln Cl - 3006/T - 0.000116R_c + 2.24t^{-0.215} \qquad (8)$$

where

Cl　=　chloride content at the bar depth
T　　=　temperature in degrees Kelvin at the bar depth
R_c　=　ohmic resistance of the cover concrete, ohms
t　　=　time in years after the initiation of corrosion.

The above relationship was determined from 2927 measurements from seven concrete chloride concentration series over five years of outdoor exposure measurements for the 3LP device. The regression coefficient and root mean square error for the above relationship are 0.95 and 0.33, respectively. No statistical model could be developed for the Gecor device because of the high variability between measured corrosion rates at a given measurement time.

Table 4 presents the 3LP corrosion rates adjusted in accordance with Equation 8. As shown, the 3LP adjusted rates are in close agreement with the weight loss measurements. The 3LP adjustment rates are about 1.55 times the weight loss measurements.

Table 4. Weight Loss and 3LP Adjusted Annual Corrosion Rate Measurements

Test Series	Exposure Period yrs	Average Annual Corrosion Rate $\mu A/cm^2$		
		Weight Loss	Adjusted 3LP	Adj/Wt.loss
18512.0	0.87	3.77	5.96	1.58
2859.6	1.84	2.34	3.58	1.53
3859.6	3.67	1.80	2.79	1.55
			Average	1.55

As shown in Table 3, the rank order of corrosion rates for the 3LP and Gecor are the same. However, the 3LP are an order of magnitude greater than the Gecor measurements. This is in agreement with the findings of other researchers. Part of the reason for the order of magnitude greater is the Tafel slope constant used by the two devices. The Gecor uses 26 mV/decade and the 3LP uses 41 mV/decade, a factor of 1.57 times greater. It is interesting that this is the same factor by which the 3LP overestimates the corrosion rate.

The remaining difference may be related to the guard ring over confining the counter electrode current because the counter electrode is much larger than the pitting corrosion area under the counter electrode. Thus, the Gecor polarized area may be much smaller than the area used in calculating the corrosion current density.

Estimating Service Life

The service of concrete structures built with bare reinforcing steel may be increased by increasing the cover depth, using a corrosion inhibitor to increase the corrosion threshold limit and/or decreasing the concrete effective diffusion constant. Table 5 presents the estimated service life using Equation 3 and adding five years for the time-to-cracking for the following conditions, 12 percentile lowest concrete cover depth = 60 and 40 mm, corrosion threshold limit 0.9 to 3.6 kg/m^3, and diffusion constants of 77, 38, and 19 mm/yr. As shown in Table 5, for service lives of 75 years and greater combination of increased cover depth, reducing the effective chloride diffusion constant and increasing the corrosion chloride threshold limit are required.

Table 5. Service Life Estimates

Corrosion kg/m^3	Cover Depth,mm	Chloride Diffusion Constant, mm^2/yr					
		77		38		19	
		40	60	40	60	40	60
0.9		14	20	20	31	30	53
1.8		19	30	29	51	49	93
2.7		28	48	47	87	84	--
3.6		48	91	89	--	--	--

Time to First Maintenance

The time to first maintenance may be estimated by using the 2.5 percentile lowest steel depth in Equation 3 and using a time-to-cracking of four years. Weyers (1994) showed that bar diameter and concrete strength parameters have little influence on the time to cracking from initiation of corrosion. Annual maintenance level can be estimated by a straight line function from the increase in damage area from 2.5 to 12 percent over the time period estimated from Equation 3.

Summary

The service life model consists of two phases, a diffusion period described by Fick's Law and a time-to-corrosion cracking from the onset of corrosion. The selection of the solution form of Fick's Law is dependent upon the surface chloride boundary condition. A constant surface chloride condition may be used for structural elements as bridge decks. A square of time function is more applicable to assessing the corrosion protection performance of the use of surface sealers and coatings. The time-to-cracking period from initiation of corrosion is strongly influenced by the rate of corrosion and is a relatively short time period of three to seven years of the structures service life.

The utility of the model lies in its ability to estimate the service of a multitude of corrosion protection methods with minimum measurable parameter requirements. Model parameters requirements are effective chloride diffusion constant, chloride corrosion threshold limit, average and standard deviation of the concrete cover depth and the surface chloride content and boundary conditions. For existing corrosion protection systems, many of these parameters have been established or can be determined from existing data. The time-to-corrosion cracking from initiation of corrosion can be estimated at about five years.

The service life model coupled with a life-cycle cost model as the one developed by Weyers, Cady and McClure (1983) may be used to determine the least cost solution to corrosion protection.

References

Cady, P. D., and R. E. Weyers (1984). "Deterioration Rates of Concrete Bridge Decks." Journal of Transportation Engineering, Vol. 110, no. 1, pp. 35-44.

Fitch, M. G., R. E. Weyers and S. D. Johnson (1995). "Deterioration of End of Functional Service Life for Concrete Bridge Decks." Transportation Research Record, No. 1490, pp. 60-66.

Liu, Y. (1996). Modeling the Time-to-Corrosion Cracking of the Cover Concrete in Chloride Contaminated Reinforcement Concrete Structures. Dissertation in Civil Engineering, Virginia Polytechnic Institute, Blacksburg, VA, p. 171.

Weyers, R. E., P. D. Cady, and R. M. McClure (1983). Cost Effective Methodology for the Rehabilitation and Replacement of Bridges. Report No. FHWA-PA-84-004, Pennsylvania Department of Transportation.

Weyers, R. E. and et. al. (1993). Concrete Bridge Protection, Repair, and Rehabilitation Relative to Reinforcement Corrosion: A Methods Application Manual. SHRP-S-360, National Research Council, Washington, DC p. 268.

Weyers, R. E. and et. al. (1994). Concrete Bridge Protection and Rehabilitation: Chemical Physical Techniques, Service Life Estimates. SHRP-S-668, National Research Council, Washington, DC, p. 357.

Zemajtis, J. and R. E. Weyers (1996). "Concrete Bridge Service Life Extension Using Sealers and Coatings in Chloride-Laden Environments." Concrete Repair, Rehabilitation, and Protection, Edited by R. K. Dhir and M. R. Jones, E & FN Spon, London, pp. 389-397.

Durability of Fiber Reinforced Plastic (FRP)
Rebars and Tendons in Aggressive Environments

Hamid Saadatmanesh[1] and Fares Tannous[2]

Abstract

This paper presents the durability test results of various commercially available composite rebars and tendons. The durability of eight different combinations of glass fiber reinforced plastic (GFRP) rebars, as well as, two carbon fiber reinforced plastic (CFRP) tendons, and one aramid fiber reinforced plastic (AFRP) tendon was examined. Tendon and rebar specimens were directly exposed to seven different solutions simulating accelerated exposure to various field conditions. Test results indicated long-term durability problems associated with GFRP rebars, while CFRP and AFRP tendons displayed excellent durability in harsh environments. The free-phase model (i.e. Fick's law) of diffusion showed to be an acceptable model to approximately predict the losses in tensile strength of composite rebars and tendons.

Introduction

The corrosion of steel reinforcement in concrete structures has been a continuous challenge facing researchers and engineers in the construction industry. Steel corrosion could result in expensive repairs or sudden failure of structures. In recent years, modern composites are becoming increasingly popular as an alternative material to steel reinforcement. By combining two or more materials with different properties, it is possible to engineer new materials such as fiber reinforced plastics with superior properties than those of the constituents. Although, FRPs have higher strength-to-weight ratio than steel, and are not subject to electrochemical corrosion, there are still some durability issues that must be considered before a widespread application of these materials. Early durability studies on composites by Shen and Springer (1976),

---

[1] Associate Professor, Department of Civil Engineering & Engineering Mechanics, The University of Arizona, Tucson, Arizona 85721

[2] Graduate Research Assistant, Department of Civil Engineering & Engineering Mechanics, The University of Arizona, Tucson, Arizona, 85721

and Loos and Springer (1979,1980) showed durability problems due to moisture absorption by glass fiber composites, and to a lesser degree in carbon fiber composites. Other studies by Uomoto and Katsuki (1995), and Zayed (1991) have reached similar conclusions, however, more investigations that would lead to better and safer designs are needed.

This paper presents analytical and experimental studies related to the accelerated exposure of non-metallic composite rebars and tendons to simulated field conditions.

Rebar and Tendon Specimens

Rebar specimens consisted of eight different combinations of GFRP rebars. Two types of glass fibers (E-glass and AR-glass), two matrix materials (polyester and vinylester), and two rebar diameters (10 mm and 19 mm) were considered. The tendon specimens consisted of two types of CFRP tendons, namely, Leadline PC-D8 8 mm diameter tendon hereafetr referred to only by "Leadline", 1x7 -7.5 mm diameter carbon fiber composite cable (CFCC) hereafter referred to only by "CFCC", and one AFRP, namely, Arapree 10 mm diameter tendon hereafter referred to by only "Arapree". GFRP rebars were pultruded from E-glass and alkali-resistant (AR) glass fibers embedded in polyester or vinylester matrix with a fiber volume fraction v_f of 72%. Leadline and CFCC tendons were fabricated from PAN-type high modulus (HM) carbon fibers embedded in epoxy matrix with a v_f of 65%. The Arapree tendon was fabricated from 350,000 Twaron HM synthetic filaments embedded in epoxy matrix with a v_f of 50%. Five specimens of each rebar type were tested in tension until failure to determine the material properties of the rebars and tendons. The initial geometric and mechanical properties for the GFRP rebars are listed in Table 1, and for the tendons are listed in Table 2, where (A_g) is the gross cross-sectional area of the rebar or tendon, (P_u) is the ultimate tensile strength, (σ_u) is the ultimate tensile stresss,(ε_u) is the ultimate strain, and (v) is Poisson's ratio. The ultimate tensile stress was calculated as (P_u / A_g) except for Arapree at which it was calculated based on the area of the fibers A_f calculated as ($v_f \times A_g$) as was suggested by Gerritse and Werner (1988).

Durability in Simulated Aggressive Environments

As previously mentioned, specimens were placed in seven different solutions that simulated conditions in the field. The solutions were 1) water H_2O at 25°C simulating exposure to rain or high humidities under bad storage conditions; 2) saturated calcium hydroxide $Ca(OH)_2$ solution with pH of 12 at 25°C simulating exposure to hydrating cement; 3) saturated $Ca(OH)_2$ solution with pH of 12 at 60°C simulating exposure to hydrating cement in hot temperatures; 4) hydrochloric acid HCl solution with pH of 3 at 25°C simulating carbonization of concrete or infiltration of

Table 1- Initial material properties of 10 mm and 19 mm diameter GFRP rebars

Material property	10 mm (E) Poly.	10 mm (E) Vinyl.	10 mm (AR) Poly.	10 mm (AR) Vinyl.	19 mm (E) Poly.	19 mm (E) Vinyl.	19 mm (AR) Poly.	19 mm (AR) Vinyl.
A_g (mm^2)	79.01	88.4	85.5	81.0	299.9	294.4	270.4	294.2
P_u (kN) S.D. (kN)	62.3 5.45	61.4 6.35	60.0 3.98	61.3 4.85	177.9 8.262	167.3 9.023	171.3 10.32	176.1 7.988
σ_u (MPa) S.D.(MPa)	788.2 22.50	694.3 11.51	701.7 46.5	756.3 59.86	593.3 11.06	568.2 12.27	633.5 38.17	598.6 27.15
E (GPa) S.D. (GPa)	55.8 2.48	56.7 2.53	53.8 5.83	54.1 5.78	56.7 2.54	53.6 2.48	53.9 5.46	51.2 5.01
ε_u (%) S.D. (%)	1.413 0.196	1.225 0.271	1.304 0.137	1.398 0.162	1.046 0.278	1.060 0.211	1.175 0.181	1.169 0.204

1 in. = 25.4 mm   1 lb. = 4.448 N   1 psi. = 6.895 kPa

Table 2- Initial material properties of CFRP and AFRP tendons

Material Property	Leadline PC-D8 8 mm φ CFRP Epoxy Matrix	1x7 CFCC 7.5 mm φ CFRP Epoxy Matrix	Arapree 10 mm φ AFRP Epoxy Matrix
A_g (mm^2)	50.3	30.4	78.6
P_u (kN) S.D. (kN)	100.5 5.120	64.9 1.121	91.2 3.794
σ_u (MPa) S.D. (MPa)	1999.2 31.7	2134.9 36.9	2324.1 48.3
E (GPa) S.D. (MPa)	149.6 12.48	146.7 5.067	120.7 15.93
ε_u (%) S.D. (%)	1.34 0.169	1.43 0.087	2.09 0.112
Poisson's Ratio (ν)	0.301	****	0.362

1 in. = 25.4 mm   1 lb. = 4.448 N   1 psi. = 6.895 kPa

acidic chemical through cracked concrete; 5) sodium chloride NaCl 3.5% by weight solution at 25°C simulating to exposure to marine environment; 6) NaCl+CaCl$_2$ (2:1) 7% by weight solution simulating exposure to NaCl and TETRA94 Anhydrous Calcium Chloride CaCl$_2$ deicing salts; and 7) NaCl+MaCl$_2$ (2:1) 7% by weight solution simulating exposure to Ice-stop Cl magnesium chloride deicing salts. In addition, 10 mm diameter GFRP rebars were subjected to ultraviolate radiation at a rate of 31.7x 10^{-6} J/sec/cm^2, and tendons were subjected to freeze-thaw cycles between temperatures of -30 and 60 °C.

Moisture Diffusion

In order to examine the effect of moisture absorption on the tensile strength of composite rebars and tendons, Fick's first law of diffusion was used to simulate the diffusion of water molecules and free ions (i.e. OH and Cl ions) into the composite specimen. Moisture diffusion would cause matrix cracking and loss in fiber area. In Fick's law or the single free-phase model of absorption in which the molecules are not bound by the material, and the driving force behind diffusion is the ion concentration (Tsai et al. 1980). To study moisture diffusion, the percentage weight gain M(%) of a composite as a function of the square root of time (t)$^{1/2}$ should be determined. To accomplish this task, rebar and tendon specimens 102 mm long with both ends epoxy-coated were oven dried until no change in weight was observed. Then they were immersed in the seven solutions , and weight changes were recorded periodically on a Mettler Balance. In order to simplify the analysis, the radius to length ratio of specimens was much less than one (i.e. r$_o$/L <<1), and the diffusivity inside the composite was assumed to be constant. Fick's law can be written as:

$$D \frac{\partial^2 C}{\partial x^2} = \frac{\partial C}{\partial t} \tag{1}$$

where "D" is the mass diffusion coefficient (i.e. diffusivity) (mm^2/min), "C" is the ion concentration (mol/l) at a distance "x" (mm) measured in a direction normal to the surface, and "t" is the exposure time (min).
A time dependent coefficient (G) can be related to the diffusivity (D) by:

$$G = \frac{m - m_i}{m_m - m_i} = 1 - \frac{8}{\pi^2} \sum_0^\infty \frac{\exp[\ (-2j+1)^2\ \pi^2\ (\frac{D\ t}{h^2})]}{(2j-1)^2} \tag{2}$$

where "h" is the specimen thickness, "m$_i$" is the initial weight of moisture in the sample, and "m$_m$" is the weight of moisture when the sample is fully saturated. Rebar and tendon samples can be considered as cylinders with radius "r", therefore, diffusion will occur in the radial direction only. Thus, the relatioship between the weight gain and the diffusivity during the initial linear stage of weight gain for a cylindrical sample is given by (Crank ,1985) :

$$\frac{m - m_i}{m_m - m_i} = 4 \sqrt{(\frac{Dt}{\pi r^2})} \tag{3}$$

In practical situations, it is of interest to determine the moisture content (m) as a percentage of the weight gain of the composite material (M). Thus, M can be defined as:

$$M = \frac{W - W_d}{W_d} \times 100 \tag{4}$$

where W and W_d are the weight of moist sample and weight of dry sample, respectively. By taking into consideration the initial moisture content (M_i) of the composite, then

$$M = G (M_m - M_i) + M_i \tag{5}$$

where M_i amd M_m are the initail percentage moisture content and percentage moisture content at saturation, respectively. From the initial linear part of the curve shown below in Fig. (1), Eq. (3) can be rearranged to obtain the diffusivity D normal to the surface as a function of the $(t)^{1/2}$ as follows:

$$D = \frac{\pi r^2}{16} (\frac{M_2 - M_1}{M_m})^2 (\frac{1}{\sqrt{t_2} - \sqrt{t_1}})^2 \tag{6}$$

where M_1 and M_2 are the percentage moisture content corresponding to times t_1 and t_2, and "r" is the radius of the test specimen.

The initial part of the curve is linear until time "t_L" indicating Fickian diffusion, then levelling off towards an asymptotic value corresponding to M_m at saturation. After extended period of time (i.e. $t > t_m$), a rapid increase in M(%) with respect to $(t)^{1/2}$ was observed indicating non-Fickian diffusion due to progressive cracking and deterioration of the matrix and fiber debonding.

To determine the extent of induced damage due to diffusion of hydroxyle (OH^-) and chloride (Cl^-) free ions in the solutions, the depth of the damaged zone "x" (mm) in which the matrix and fibers are considered to be ineffective in transferring tensile

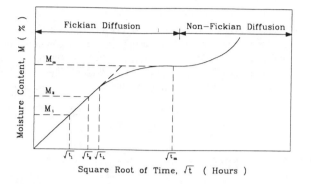

Figure 1. Typical M(%) versus $(t)^{1/2}$ relationship for FRP rebars and tendons

forces is calculated as follows:

$$x = \sqrt{2 \cdot D \cdot C \cdot t} \tag{7}$$

where "C" is the hydroxyl and chloride ion concentrations in (mol/l). If the initial radius of the specimen is "r_o" (mm), the radius of the residual area is "r_r" (mm) as shown in Fig. 2, and the initial tensile strength is P_u, then the predicted residual tensile strength P_p is calculated from:

$$P_p = P_u \left(1 - \frac{x}{r_o} \right)^2 = P_u \left(\frac{r_r}{r_o} \right)^2 \tag{8}$$

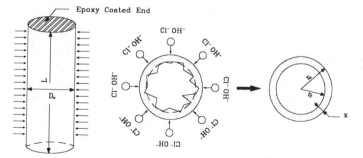

Figure 2. Diffusion of (OH⁻) and (Cl⁻) free ions into FRP rebars and tendons

The two-phase diffusion model has also been used to determine the diffusivity and weight gain in composites when exposed to temperature and humidity (Gurtin and Yatomi, 1979) and (Lo et al., 1982). It is defined by introducing the following two parameters: 1) the time rate of moisture desorption "b"; and 2) the time rate of moisture absorption "a". The time dependent coefficient (G) for the two-phase model is given by:

$$G = 1 - (\frac{a}{a+b}) \exp [-bt] - (\frac{b}{a+b}) \sum_{j=0}^{\infty} [\frac{1}{(2j+1)^2} \exp (-\frac{\pi^2 D \ t}{r^2} (2j+1)^2)] \qquad (9)$$

For a suffuciently long period of time, G can be approximated by:

$$G = 1 - (\frac{a}{a+b}) e^{-bt} \qquad (10)$$

Equation (10) lists the relationship between the time dependent coefficient (G) and time (t). Taking the natural logarithm of Eq. (10) and rearranging the terms, the follwing is obtained:

$$\ln (1- G) = \ln (\frac{a}{a+b}) - bt \qquad (11)$$

Parameters "a" and "b" can be determined from Eq. (11), where "b" is the slope and "a" is the zero time intercept. The diffusivity from the two-phase diffusion model (D*) can be related to the free-phase model diffusivity (D) through the following equation:

$$D^{\cdot} = D^{\cdot}(\frac{a+b}{b}) \qquad (12)$$

Results and Discussion

Rebar and tendon specimens were directly exposed to seven different solutions for six months, and then tested in tension until failure and changes in the mechanical properties were recorded. Furthermore, the effects of exposure to ultraviolet radiation and freeze-thaw cycles on the mechanical properties of rebars and tendons was also examined. Tension test results of 10 mm and 19 mm diameter GFRP rebars are listed in Tables 3, and for FRP tendons are listed in Table 4.

Durability in Water (H_2O), T= 25 °C

Generally, losses of tensile strength of FRP rebars and tendons were the lowest. Fick's law was applicable for calculating the diffusivity , but Eqs. (7) and (8) could not

Table 3 - Test results of 10 mm and 19 mm diameter GFRP rebars in solutions

Environment	D x10⁷ mm²/ min	C mol /l	t x 10⁻⁵ min	x mm	E GPa	E_r GPa	P_u kN	P_r kN	P_p kN	P_r/P_p (%)
10mm φ E- P										
pH= 12, T= 25 °C	3.5	1.04	2.59	0.44	55.8	54.3	62.3	46.7	51.9	0.90
pH= 12, T= 60 °C	5.5	1.04	2.59	0.55	55.8	56.1	62.3	44.5	49.5	0.90
NaCl 3.5%	4.4	0.60	2.59	0.49	55.8	51.3	62.3	55.5	53.3	1.04
10mm φ E- V										
pH= 12, T= 25 °C	1.3	1.04	2.59	0.27	56.7	55.4	61.4	53.4	55.4	0.96
pH= 12, T= 60 °C	1.5	1.04	2.59	0.29	56.7	52.7	61.4	48.9	55.0	0.89
NaCl 3.5%	1.4	0.60	2.59	0.21	56.7	58.1	61.4	57.4	56.7	1.01
10mm φ AR- P										
pH= 12, T= 25 °C	3.1	1.04	2.59	0.41	53.8	51.7	60.0	47.5	50.9	0.93
pH= 12, T= 60 °C	5.5	1.04	2.59	0.53	53.8	53.9	60.0	43.2	48.1	0.90
NaCl 3.5%	4.2	0.60	2.59	0.36	53.8	53.3	60.0	50.1	52.0	0.96
10mm φ AR- V										
pH= 12, T= 25 °C	1.2	1.04	2.59	0.26	54.1	54.4	61.3	53.5	55.3	0.97
pH= 12, T= 60 °C	1.7	1.04	2.59	0.30	54.1	53.3	61.3	47.4	54.2	0.88
NaCl 3.5%	1.1	0.60	2.59	0.18	54.1	56.7	61.3	52.3	57.0	0.92
19mm φ E- P										
pH= 12, T= 25 °C	3.9	1.04	2.59	0.46	56.7	51.7	178	145	162	0.90
pH= 12, T= 60 °C	8.2	1.04	2.59	0.67	56.7	54.7	178	142	155	0.92
NaCl 3.5%	7.8	0.60	2.59	0.49	56.7	55.9	178	164	160	1.03
19mm φ E- V										
pH= 12, T= 25 °C	1.5	1.04	2.59	0.29	53.6	51.3	167	149	158	0.94
pH= 12, T= 60 °C	1.9	1.04	2.59	0.32	53.6	53.4	167	147	156	0.94
NaCl 3.5%	4.6	0.60	2.59	0.38	53.6	55.2	167	160	154	1.04
19mm φ AR- P										
pH= 12, T= 25 °C	4.5	1.04	2.59	0.49	53.9	56.4	171	149	154	0.97
pH= 12, T= 60 °C	9.8	1.04	2.59	0.73	53.9	54.3	171	142	146	0.97
NaCl 3.5%	9.2	0.60	2.59	0.54	53.9	58.2	171	160	152	1.05
19mm φ AR- V										
pH= 12, T= 25 °C	2.3	1.04	2.59	0.35	51.2	52.3	176	159	164	0.97
pH= 12, T= 60 °C	3.1	1.04	2.59	0.41	51.2	50.2	176	151	162	0.93
NaCl 3.5%	5.3	0.60	2.59	0.40	51.2	54.7	176	165	162	1.02

1 in. = 25.4 mm   1 lb. = 4.448 N   1 psi. = 6.895 kPa

Table 4 - Test results of Leadline, CFCC, and Arapree FRP tendons in solutions

Environment	D $\times 10^8$ mm²/min	C mol/l	$t \times 10^{-5}$ min	x mm	E GPa	E_r GPa	P_u kN	P_i kN	P_p kN	P_p/P_i (%)
Leadline PC-D8										
pH= 12, T= 25 °C	0.15	1.04	5.18	0.04	150	151	101	97.1	98.5	98.6
pH= 12, T= 60 °C	0.54	1.04	5.18	0.08	150	151	101	96.6	96.7	99.9
NaCl 3.5%	0.88	0.60	5.18	0.07	150	150	101	95.3	96.8	98.5
1x7 - CFCC										
pH= 12, T= 25 °C	4.60	1.04	5.18	0.22	147	141	64.9	64.3	55.9	115
pH= 12, T= 60 °C	6.90	1.04	5.18	0.09	147	146	64.9	64.1	54.0	119
NaCl 3.5%	13.0	0.60	5.18	0.28	147	151	64.9	64.5	53.6	120
Arapree 10 mm										
pH= 12, T= 25 °C	2.80	1.04	5.18	0.17	121	115	91.2	87.3	85.0	103
pH= 12, T= 60 °C	4.40	1.04	5.18	0.22	121	110	91.2	85.4	83.4	102
NaCl 3.5%	1.80	0.60	5.18	0.11	121	112	91.2	87.3	87.4	99.9

1 in. = 25.4 mm   1 lb = 4.448 N   1 psi = 6.895 kPa

be used to estimate the losses in the tensile strength since the ion concentration "C" is equal to zero, thus resulting in a moisture penetration of zero (i.e. $x = 0$). For example, the reduction in tensile strength of 10 mm AR-glass/polyester and vinylester rebars was 7.3% and 5.8%, respectively. Limited changes in the elastic modulus were observed and were withinm 5% of the original. Generally, a reduction in the ultimate strian was observed and it was higher for 10 mm diameter rebars than for the 19 mm diameter rebars. As for FRP tendons, the diffusivity D and maximum moisture content M_m were lowest for Leadline and CFCC in water as compared to other solutions. The diffusivity and M_m values were overestimated for CFCC. This was due to the surface texture of CFCC which allowed moisture to infiltrate between the gaps and cavities created by the carbon thread wound around each one of the 2.5 mm individual strands. Test results indicate that water has limited effect on the tensile strength of all three tendons. For eaxmple, the loss of tensile strength for Leadline after one year was 0.7%, for CFCC it was 0.31%, and for Arapree it was 2.3%. Limited changes in E, ε_u, and v were observed, thus indicating no changes in material properties.

Durability in Alkaline Solution (pH=12), T= 25 and 60 °C

In all cases, the diffusivity increased with the increase in temperature indicating a direct relationship between temperature and diffusivity. For E-glass rebars, losses in tensile strength was among the highest in alkaline solution. Lower diffusivity was recorded in vinylester rebars than in polyester ones. This was reflected in lower depth of moisture penetration "x" and losses in tensile strength as indicated by Eqs. (7), and (8), repectively. For example, in the case of alkaline solution at temperatures of 25 and 60 °C, the measured losses in strength of 10 mm polyester rebars were 25.0 and 28.6%, respectively, as compared to 10 mm vinylester rebars were the measured losses were 13.0 and 20.3%, respectively. Similar observations were recorded for 20 mm rebars, however, the recorded percentage losses were lower because of the larger percentage of residual area of undamaged fibers and matrix.

Similar observations to those of E-glass rebars were recorded for AR-glass rebars. The diffusivity of AR-glass/polyester rebars was lower than the vinylester ones, and it increased with the increase in temperature. The highest percentage of loss in tensile strength after 6 months was recorded in alkaline solution at 60 °C. For the 10 mm polyester rebars it was 28.0%, and for vinylester it was 22.6%. As for the 19 mm polyester rebars it was 16.9%, and 14.1% for the vinylester ones. The difference between the actual residual tensile strength (P_r) and the predicted strength (P_p) were less than 13% for E-glass rebars and within 10% for the AR-glass rebars. In addition, better predictions were obtained for 19 mm rebars than for 10 mm rebars. This indicates that Fick's law is an acceptable method to predict the loss of strength due to diffusion in alkaline environment.

Test results showed little effect of alkaline solution on changes in the ultimate tensile strength, elastic modulus, and Poisson's ratio of Leadline and CFCC in alkaline solutions. This clearly demonstrates that PAN-type carbon fibers in conjunction with epoxy matrix possess excellent durability in alkaline environment (i.e. concrete). The maximum moisture content at saturation (M_m) at 25 °C was lower than that at 60 °C for all three types of tendons as listed in Table 4. Good agreement between the measured residual strength and the predicted residual strength after 6 months and 12 months was observed, thus Fick's law can adequately predict changes in material properties of Leadline and Arapree. However, for CFCC the residual tensile strength was underestimated. For example, the residual strength of CFCC after 12 months in alkaline solution at 60°C was 98.8% of ultimate, while the predicted strength using Fick's law was 83.2% of ultimate. This is due the overestimated diffusivity (D) due to the moisture and $Ca(OH)_2$ residue depositing inside the gaps and cavities created by the carbon threads wrapped around the carbon wires, not as a result of OH^- ions diffusion into the composite. Visual inspection of CFCC specimens showed a white residue (i.e. $Ca(OH)_2$) formation on the surface. This case demonstrates Fick's law inability to take into account the other factors such as surface texture that would reduce the effectiveness of this approach to obtain reliable results. Another factor that could contribute to the underestimation of the redidual strength is that the fibers in the contaminated zone could still contribute to the load carrying capacity of the tendon since the chemical resistance of carbon fibers is very good. This however is not the case for E-glass and AR-glass rebars, dut to the poor chemical resistance to alkali attack. In this case fibers in the contaminated zone lost their load carrying capacity, thus resulting in better agreement between the predicted and measured strength.

Durability in Acidic Solution (pH=3), T=25°C

In the case of acidic solution with pH of 3, the reported losses in strength were lower than in the case of alkali or salt solutions. This resulted from the low concentration of Cl^- ions, thus leading to lower computed "x" values and no noticeable loss in tensile strength. This indicates that other factors could play a role in strength loss such as water and hydrogen ion H^+ diffusion. In general, Fick's law predicted losses of strength within 15% of the measured values. The diffusivity and maximum moisture content for tendons were the lowest for Leadline. No noticeable loss of tensile strength was observed, and good agreement between the measured and predicted strength was obtained. As for CFCC, the moisture diffusion was initially Fickian, however, it became non-Fickian after only 100 hours. The close agreement between the measured and predicted loss in this case is just coincidental. As for Arapree, a 8.3% loss in tensile strength was recorded after 12 months of exposure. Moreover, a progressive reduction in Poisson's ratio and elastic modulus with respect to time was also observed. The predicted loss in tensile strength was higher than the measured loss of strength. As previously stated, this is due to the fact that fibers in the contaminated zone are not totally destroyed, thus resulting in higher residual strength than predicted.

Durability in Sea Water and Deicing Salt Solutions, T=25°C

Lower losses in tensile strength were measured in sea water than in deicing salt solutions, in spite of close diffusivities. This is due to higher concentrations of Cl^- present in the solution. For 10 mm diameter E-glass/polyester and vinylester rebars in sea water, losses in tensile strength were 10.9 and 6.5%, respectively. Similarly for 19 mm diamter E-glass/polyester and vinylester rebras, losses in sea water were 8.3% and 4.6 %, respectively. Higher losses were reported in deicing salt solutions for 10 mm E-glass/polyester and vinylester rebars. The measured losses were 26.7 and 22.9%, respectively. This raises durability concerns regarding the prolonged exposure of E-glass rebars exposed to sea water and deicing salts. Similar observations could be also made concerning AR-glass rebars in sea water and deicing salt solutions. In general, Fick's law was an adequate approach to estimate losses of strength for E-glass and AR-glass rebars exposed to sea water and salt solutions.

For Leadline, the measured M_m and diffusivity were relatively insensitive to the concentration or type of solution. A relatively small reduction in the tensile strength was observed for specimens in sea water or salt solutions. For example, 5.2% loss in tensile strength of Leadline was recorded after one year. Limitted changes in the elastic modulus or Poisson's ratio was observed after 12 months of exposure. Non-Fickian diffusion was observed in the case of CFCC cable due to the same aforementioned reasons. Limited losses in tensile strength or changes in the elastic modulus were observed due to the excellent ability of the PAN-type carbon to resist chloride attack. Fickian diffusion was exhibited by Arapree and salt solutions. The diffusivity was little affected by the salt type or concentration. Moreover, the diffusivity and M_m was higher in acidic and salt solutions than in alkaline solutions. this indicates that chloride ions has greater ability than hydroxyl ions to penetrate (i.e. diffuse) into aramid composites.

In general, the two-phase model provided more accurate predictions than the free-phase model, specially in the transitional zone between the initial linear part of the curve and the part asymptotic to M_m. Furthermore, in all cases the rate of moisture absorptoion was much lower than the rate of moiture desorption, thus the free-phase and the two-phase diffusivities were close, and after a sufficiently long period of time, the free-phase model and the two-phase model would converge (Tannous, 1997). It is important to point out that samples in direct exposre to alkaline and salt solutions exaggerates actual exposure in the field. For example, alkaline solution rich with OH⁻ free ions could accurately reflect the effect of freshly poured concrete where pores contain free OH⁻ ions, but after relatively short time, and as a result of hydration, OH⁻ ions are trapped within the hardened gel which greatly reduces their chemical reactivity. Considering the slow nature of diffusion and the existence of barriers that would furthermore slow it down such as concrete cover and dry weather, it would take longer period of time in the field to induce the same results obtained in the lab. Further studies are needed to extrapolate accelerated test results to predict the long-term durability of composites under field conditions.

Durability Under Ultraviolet Radiation

Exposure to ultraviolet light caused the color of specimens to darken, but no significant changes in material propertie was observed. Estimated losses in tensile strength were less than 3%. The total energy absorbed by the rebars over a six month period was estimated at 0.5 kJ/cm².

Durability Under Freeze-Thaw Cycles

No measurable changes in material properties of Leadline, CFCC, and Arapree were recorded after 1200 freeze-thaw cycles between -30 and 60℃. Test results indicate that under dry conditions, freeze-thaw cycles have limited to no effect on the material properties. Induced thermal strains during hot and cold cycles were not significant enough to induce thermal fatigue. Furthermore, temperature fluctuations between hot and cold cycles were applied gradually (i.e. similar to field conditions), which eliminated the possible effect of thermal spikes in inducing thermal cracking. As a result, thermal strains were well within the elastic range for the tendons. A larger number of cycles (i.e. > 1200 cycles) is required to induce thermal fatigue damage. For example, the combined exposure to freeze-thaw cycles and deicing salts could more accurately reflect field conditions.

The effect of freeze-thaw cycles is most pronounced when the fiber and matrix are thermally incompatibel. Although, freeze-thaw action may not be enough to induce measurable amounts of damage through matrix cracking and fiber debonding, it may facilitate (i.e. accelerate) the diffusion of Cl^-, OH^- ions and other contaminents, thus leading more aggressive and serious forms of degradation.

Conclusions

1. The diffusivity is dependent on temperature and the type of solution.
2. Under most conditions, Fick's law may be used to predict changes in moisture content and mechanical properties with respect to time until M_m is reached.
3. Vinylester has lower diffusivity and better resistance to chemical attack than polyester.
4. E-glass and AR-glass displayed durability problems under accelerated exposure to alkali attack, marine environment, and deicing salts, and it was more pronounced for smaller diameter rebars.
5. Leadline, CFCC, and Arapree exhibited excellent resistance to alkali and salt attack.
6. Fick's law was not applicable in the case of CFCC cable due to non-Fickian diffusion induced by the surface texture.
7. Ultraviolet light has limited effect on the tensile strength of GFRP rebars.
8. Freeze-thaw cycles has limited effects on the durability of carbon and aramid tendons.
9. In most cases, non-fickian diffusion showed by the rapid increase in moisture content was obserevd beyond saturation as a result of excessive cracking of matrix and fiber debonding.

Acknowledgement

This work was sponsored by The National Science Foundation under Grant No. MSS-9257344, and The Federal Highway Administration under Grant No. DDEGRF-93-03. The Support of The National Science Foundation and the Federal Highway Administration is greatly appreciated.

APPENDIX

References

Crank, C., "Mathematics of Diffusion," Clarendon Press, Oxford, 1985.

Gurtin, M.E., and Yatomi, C., "On a Model for Two-Phase Diffusion in Composite Materials," J. Composite Materials, Vol. 13, April 1979, pp. 126-130.

Gerritse, A., and Werner, Jürgen, "Arapree, The Prestressing Element Composed of Bonded Twaron Fiber," Technical Data Report, Sireg Corporation, Italy, Sept. 1988, pp. 3-6.

Katsuki, F., and Uomoto, T., "Prediction of Deterioration of FRP Rods Due to Alkali Attack," Non-Metallic FRP Reinforcement for Concrete Structures, Proceedings of the Second Intl. RILEM Symposium (FRPCS-2), Aug. 1995, pp. 82-89.

Lo, S. H., Hahn, H. T., and Chiao, T. T., "Swelling of Kevlar 49/Epoxy and S-2 Glass/Epoxy Composites," Progress in Science and Engineering of Composites Vol. 2, Proceedings of the Fourth International Conference on Composites, Tokyo, Japan 1982, pp. 987-1000.

Loss, A. C., and Springer, G. S., "Moisture Absorption of E-glass Composites," J. Composite Materials, Vol. 14, April 1980, pp. 142-153.

Shen, C. H., and Springer, G. S., "Moisture Absorption and Desorption of Composite Materials," J. Composite Materials, Vol. 10, Jan. 1976, pp. 2-20.

Tannous, F. E., "Durability of Non-Metallic Reinforcing Bars and Prestressing Tendons," A Dissertation Submitted to the Faculty of the Department of Civil Engineering and Engineering Mechanics, University of Arizona, Tucson, Arizona 1997.

Tsai, S. T., and Hahn, H. T., "Introduction to Composite Materials," Technomic Publishing Co., Inc., 1980, pp. 334-342.

Zayed, A. H., "Deterioration Assessment of Fiber-Glass Plastic Rebars in Different Environments," The NACE Annual Conference and Corrosion Show, Paper No. 130, March 1991, Cincinnati, Ohio.

NEW APPROACH FOR TENSILE TESTING OF FRP COMPOSITES

Houssam A. Toutanji[1] and Tahar El-Korchi[2]
Associate Member, ASCE

Abstract

The use of fiber reinforced polymer composite (FRPC) wrap is increasingly being used for rehabilitation and retrofit of cement and concrete structural elements. As a consequence an accurate uniaxial tensile strength data are required for reliable design with these composites. A new room temperature tensile test technique can contribute to attaining this goal. The technique minimizes errors due to gripping and load train misalignment often encountered in testing brittle materials and utilizes a specimen with a simple geometry. This paper describes this simple technique which utilizes the ASCERA hydraulic tensile tester.

Three types of fiber composite wraps including two carbon fibers and one glass fiber were applied with a two-part epoxy resin system. Tensile strength values of the carbon fiber wrapped cement composites were in the range of 67 to 125 MPa. The glass fiber wrapped composites were in the range of 33 to 59 MPa. The tensile strength of the unwrapped cement-based specimens were under 8 MPa. The tensile strength data of the fiber with and without resin-matrix were obtained using the law of mixtures and constitutive models.

Introduction

Advanced composite materials such as fiber reinforced polymer (FRP) have the potential to revolutionize engineering technology. But in order to use these advanced composite materials, a detailed knowledge of their mechanical behavior is imperative. In addition, structurally efficient design using these advanced composites mandates a detailed knowledge of their mechanical properties. Some of the mechanical characterization of these FRP composites has been recently studied, and is currently undergoing further investigation at ambient room temperature.

---

[1]Associate Prof., Dept. of Civil Engineering, University of Puerto Rico, Mayaguez, PR 00681.
[2]Associate Prof., Dept. of Civil and Environmental Eng., Worcester Polytechnic Institute, Worcester, MA 01609.

However, research on their tensile strength characteristics is still limited. The tensile strength of FRP composites is the most important property which gives the strengthened or repaired structural element its new improved characteristics, but yet reliable data on the tensile properties of these materials are still lacking.

It is well known that it is very difficult to conduct stable tests under uniaxial tension. These difficulties arise from two areas: (1) stress concentration at gripping and (2) amount of eccentricity. These factors may significantly influence the test results, producing measurements that underestimate the intrinsic tensile strength of the material. Very small eccentricity in loading, for example 5-10%, could result in a load reduction of 25-50% in calculated direct tensile strength (John and Shah 1989).

The past decade has seen the emergence of several new and improved tensile testing procedures with incorporated improved gripping and alignment procedures, and are easier to use than past methods (Amaral and Pollock 1989). However, most of these new tensile test procedures are designed for high temperature testing (Liu et al. 1989; Soma et al. 1986).

Baratta and Driscoll (Driscoll and Baratta 1971) developed a simple, self-aligning hydraulic tensile tester for room temperature use, which significantly reduced errors due to alignment, bending or gripping. However, this technique utilized a cylindrical "dogbone" shaped specimen which was expensive to machine. Researchers in Sweden (Hermanson et al. 1987), improved on Baratta and Driscoll's concept so that a simple cylindrical shape specimen can be utilized. The improved technique utilizes the ASCERA hydraulic tensile tester, manufactured by a Swedish engineering firm.

This technique has since been used to evaluate the tensile strength for a number of structural materials (Toutanji and El-Korchi 1994; Katz et al. 1993; Lucas 1991), such as: cement-based composites, silica-fume cementitious composites, carbon fiber-reinforced cementitious composites, graphite, and high strength ceramics.

Test results showed that the uniaxial tensile strength of cementitious composites tested using the ASCERA hydraulic tensile tester is higher than those obtained using the traditional uniaxial tensile test (Toutanji 1992). This is basically due to the minimization of load eccentricity and gripping effect inherent in traditional uniaxial tensile tests. The use of the hydraulic tensile technique provides a number of advantages, including minimization of eccentricity and excessive gripping stresses, identification of the origin of fracture, and enables testing of relatively large volumes with small size specimens.

The purpose of the present paper is to use this method to evaluate the uniaxial tensile strength of FRP composite sheets and cement-based specimens wrapped with FRP composites. Based on the obtained results of the cementitious wrapped specimens, the tensile strength of the fiber sheets is theoretically calculated using the rule of mixtures.

Experimental Procedure

Materials

The cementitious specimens were made of silica-fume cement paste in which 8% of cement by mass was replaced by silica fume. The water-cementitious materials ratio was 0.30. ASTM Type II portland cement was used. Superplasticizer was added as a percentage of the total weight of the cementitious materials (portland cement and silica-fume).

Three types of FRP tow sheets were used: two carbon (C1 and C5) and one glass (GE). The two-part resin system used consists of resin and hardener mixed in the ratio of 2:1. A summary of some of the properties of the FRP sheets and the epoxy is included in Table 1.

TABLE 1.   Mechanical properties of FRP sheets and epoxy (reported by the manufacturer)

FRP sheet	Tensile strength (MPa)	E (GPa)	Ultimate strain (%)	Thickness (mm)
GE	1,518	69	2.1	0.118
C1	3,485	228	1.5	0.110
C5	2,938	373	0.8	0.165
Epoxy	55.90	2.35	2.4	–

Tensile Testing Technique

The tensile test is a self-alignment hydraulic technique that utilizes a hydraulic tension tester manufactured by ASEA®, an engineering firm located in Sweden. The tensile specimens are cylindrical bars measured 16 mm in diameter and 120 mm in length. Forty mm on each end of the cylindrical bar is inserted into a steel piston and adhesively bonded in place with high-strength epoxy. The specimen-piston assembly is inserted into a high pressure chamber. The pressure is applied and increased until the specimen is broken apart by the hydraulic pressure acting against the pistons. The pressure chamber, with specimens inserted, is shown schematically in Fig. 1. The internal pressurization of the specimen suspended between the O-ring seals minimizes bending stresses. The nominal tensile fracture stress may be calculated by using the hydraulic pressure at failure and the geometric parameters of the specimen-piston assembly. A small degree of triaxiality in the hydrostatic loading of the specimen exists causing a small deviation from the uniaxial stress state, calculated, using the Coulom-Mohr theory of fracture, to be less than 5% (Toutanji 1992). In addition, to eliminate the effects of stress intensification at the specimen to piston bond transition, data from specimens that fractured within one half the radius of the specimen from the piston glue line are not used. Fractures occurring in these locations are considered invalid tests (ASEA CERAMA AB 1988).

The stress is related to the hydraulic pressure and test geometry by the following expression:

$$\sigma = \left(\frac{A - A_S}{A_S} \right) x \ P$$

(1)

Where σ is the fracture stress, A is the cross-sectional area of the piston, A_s is the cross sectional area of the specimen, and P is the pressure at failure. The correction for eccentricity in this study was minimal, since the specimens were straight and thus didn't exhibit any significant bending stresses.

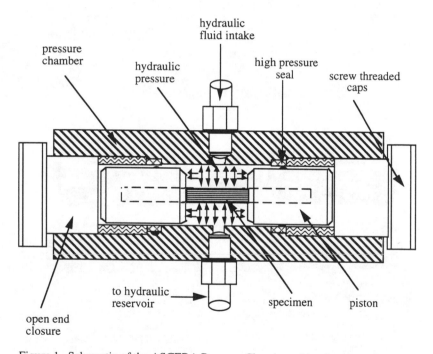

Figure 1. Schematic of the ASCERA Pressure Chamber with a Specimen Inserted

Specimen Preparation

The specimens consisted of small bar specimens measured 16 mm in diameter and 120 mm in length. The specimens were kept in their molds sealed with plastic sheets for at least 24 hours to prevent moisture evaporation. Specimens were moist cured for 56 days at a temperature of +25°C and a relative humidity in excess of 90%.

The cementitious bar specimens were cleaned and surface dried before the epoxy was applied. A thin layer of epoxy was applied on the surface of the samples. An unidirectional FRP sheet was then applied directly to the surface. Slight finger pressure was applied on the surface to ensure that there was no air trapped between the FRP sheet and the cementitious sample. Only one lap of the FRP sheet was applied. The fibers were placed along the longitudinal direction of the specimen, perpendicular to its diameter. All wrapped and unwrapped specimens were tested using the ASCERA hydraulic tension tester. The operating procedure of the tester has been described (ASEA CERAMA AB 1988) and was closely followed in testing the specimens.

FRP composite sheets were also prepared and tested using the ASCERA technique. The FRP composite specimens were wrapped along two pieces of cement paste samples at the edge with an empty space in the center. When the specimen was wrapped, a hollow FRP cylinder was formed and the filled ends were inserted into the pistons, see Fig. 2.

A hollow FRP cylinder

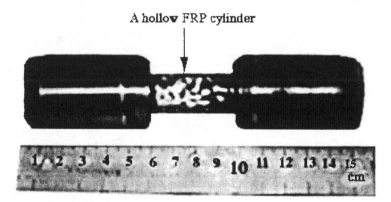

Figure 2. FRP ASCERA Specimen, the FRP Specimen is Hollow in the Center

The FRP composite sheets were also tested using a traditional uniaxial tension tester. Thus the results obtained by ASCERA were compared to those obtained by the traditional uniaxial tensile test. The specimens were prepared and tested in accordance with the guidelines of ASTM 3039.

Results

Tensile Strength of FRP Composite Sheets

The tensile strength results of the FRP composite sheets with and without resin-matrix measured by the traditional uniaxial tensile technique (UTT) are shown in Table 2. The tensile strength values of fiber without resin matrix were designated as UTT-NR and those with resin-matrix were designated as UTT-WR. Each of the UTT-NR and UTT-WR strength values represent the average strength of six specimens.

TABLE 2. Tensile strength results obtained using the uniaxial tensile test (MPa)

Specimen	UTT-NR	UTT-WR
GE	1,007	1,453
C1	2,240	3,785
C5	2,112	2,757

Tensile Strength of Cement-based FRP Wrapped Specimens

The tensile strength values of the cementitious specimens wrapped with each of the GE, C1, and C5 fiber sheets are shown in Fig. 3. The number of specimens, average tensile strength, and the standard deviation for each composite are provided on the bar charts. All specimens were tested using the ASCERA hydraulic tester. The cement-based specimens wrapped with each of the FRP sheets, C1, C5, and GE, were designated as TC1, TC5, and TGE, respectively. Specimens wrapped with C5 fiber exhibited the highest strength and specimens wrapped with the GE fiber exhibited the least strength. The average tensile strength of TGE specimens was 45 MPa, TC1 was 76 MPa, and TC5 was 98 MPa. Based on these strength values, the tensile strength of the fiber sheets can be back calculated using the Law of Mixtures. The Law of Mixtures equation can be written as follows (Balaguru and Shah 1992):

$$\sigma_{cu} = V_m\sigma_{mu} + V_f\sigma_{fu} + (1-V_m-V_r)\sigma_{ju} \tag{2}$$

Where:

σ_{cu} = composite tensile strength

σ_{mu} = matrix tensile strength

σ_{fu} = fiber tensile strength

σ_{ju} = resin matrix tensile strength

V_m = matrix volume fraction

V_f = fiber volume fraction

Since the average cross-sectional area of the cement-based FRP wrapped specimen, the cross-sectional area of the cement-based specimen, and the thickness of the FRP sheets are known, the thickness of the epoxy can be estimated. Hence, the volume fraction of each of the matrix, fiber, and epoxy can be easily calculated. Therefore, the only unknown in Eq. 2 is the tensile strength of the fiber.

Substituting σ_{cu}, v_m, σ_{mu}, v_f, σ_{gu} in the Law of Mixtures equation, the strength of the fiber σ_{fu} (fiber alone without resin matrix) can be evaluated.

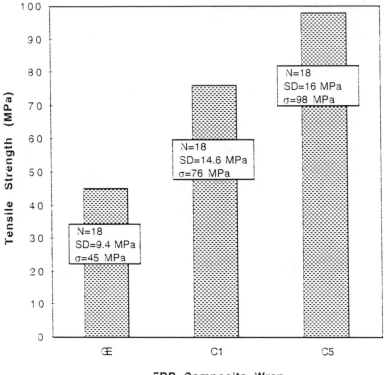

Figure 3. Tensile Strength of Specimens Wrapped with FRP Composite Sheets

The fiber sheets with resin-matrix were tested using the ASCERA technique; strength values are shown in Table 3. The FRP composite specimens were wrapped along two pieces of cement paste samples at the edge with an empty space in the center.

TABLE 3. Tensile Strength results obtain using the ASCERA technique

Specimen	Wrapped cement-based specimen (MPa)	Fiber strength-NR (MPa)	Fiber strength-WR (MPa)
TGE	45	1,217	1,711
TC1	76	2,564	3,938
TC5	98	2,340	2,940

Comparison Between UTT and ASCERA

Figure 4 shows the tensile strength results obtained by UTT and ASCERA techniques for both FRP sheets with resin-matrix and without resin-matrix.

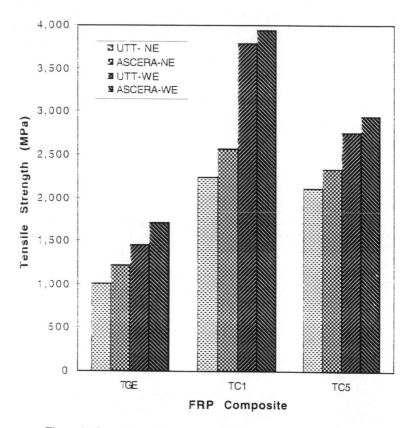

Figure 4. Comparison Between the UTT and ASCERA Methods

In the case of FRP sheets with no resin-matrix, the tensile strength obtained by ASCERA for each of the GE, C1, and C5 fiber sheets were 21, 14, and 11% higher than the strength values obtained by UTT, respectively. In the case of FRP sheets with resin-matrix, the tensile strength obtained by ASCERA for each of the GE, C1, and C5 fiber sheets were 18, 4, and 7% higher than the strength values obtained by UTT. The higher strength values obtained by ASCERA as compared to UTT is due to the fact that ASCERA technique provides an uniform stress distribution throughout the length of the specimen and thus minimizes stress concentration at gripping and misalignment. The strength difference between the two methods seemed to be less for specimens with resin-matrix. This may be attributed to the fact that the resin-matrix in the fiber provides better alignment during testing.

Conclusions

The following conclusions can be drawn from this research:
1. The tensile strength test of cement wrapped specimens with fiber composite sheets can be used as a good test for calculating the fiber strength, composite effect, and cement-epoxy compatibility.
2. The uniaxial tensile strength values of the FRP composites with and without resin-matrix obtained using the ASCERA hydraulic tensile tester are up to 20% higher than those obtained by the traditional uniaxial tensile technique.
3. The difference in strength obtained from the two techniques (UTT and ASCERA) decreases for specimen with resin-matrix as compared with those of fiber alone. This may be attributed to the fact that the resin-matrix in the fiber provides better alignment during testing.
4. The presence of resin matrix in the fiber sheets significantly increases the tensile strength.

Acknowledgment

The authors would like to acknowledge the financial support of the National Science Foundation CAREER Grant CMS-9501541. The contribution of materials from Tonen Inc. is also gratefully acknowledged.

References

Amaral, J. E. and Pollock, C. N. (1989), *Mechanical Testing of Engineering Ceramics at High Temperature*, eds. B.F Dyson, R.D. Lohr, and R. Morrell, Elsevier Applied Science, London and New York, pp. 51-68.

ASEA CERAMA AB (1988), "Introduction and Operation Manual for ASCERA Hydraulic Tensile Testing," Robertsfors, Sweden.

Balaguru, P., Shah, S. (1991), *Fiber Reinforced Cement Composites*, McGraw Hill, NY, 53pp.

Driscoll, G. W., and Baratta, F. I. (1971), "Modification to an Axial Tension Tester for Brittle Materials" Army Materials and Mechanics Research Center, AMMRC TR 71-3, Watertown, MA.

Hermanson, L., Adlerborn, J., and Burstrom, M. (1987), "Tensile Testing of Ceramic Materials," *High Tech Ceramics*, ed. P. Vincenzini, Elsevier Science Publishers B.V., Amsterdam, pp. 1161-1168.

John, R. and Shah, S. P. (1989), "Fracture Mechanics Analysis of High-Strength Concrete," *ASCE Materials Journal*, Vol. 1, No. 4, pp. 185-197.

Katz, R. N., Wechsler, G., Toutanji, H. A., Freil, D., Leatherman, G. L., and Rafaniello, W. (1993), "Room Temperature Tensile Strength of AlN," *Proceedings, Ceramic Engineering & Science*, Vol. 14, No. 718, pp. 282-291.

Liu, K. C., Pih, H., and Voorhes, W. (1989), *Mechanical Testing of Engineering Ceramics at High Temperature*, eds. B. F. Dyson, R. D. Lohr, and R. Morrell, Elsevier Applied Science, London and New York, pp. 69-87.

Lucas, P. H. (1991) "Direct Tensile Testing of Brittle Materials," Masters Thesis, Worcester Polytechnic Institute, MA.

Soma T., Matsui M., and Oda, I. (1986), *Non-Oxide Technical and Engineering Ceramics*, ed. S. Hampshire, Elsevier Applied Science, London and New York, pp. 361-374.

Toutanji, H. A. (1992), "The Development of a Cementitious Composites Axial Tensile Technique and its Application to Carbon Fiber-Reinforced Cementitious Composites," Ph.D. Thesis, Worcester Polytechnic Institute, Worcester, MA.

Toutanji, H. A., and El-Korchi, T. (1994), "Uniaxial Tensile Strength of Cementitious Composites," *Journal of Testing and Evaluation*, JTEVA, Vol. 22, No. 3, pp. 226-232.

Strengthening of RC Flexural Members with FRP Composites

Antonio Nanni[1] and William Gold[2]

Abstract

Experimental data obtained from strengthened, pre-cracked, reinforced concrete (RC) specimens are presented. Strengthening was attained by the adhesion of carbon fiber reinforced plastic (CFRP) sheets to the concrete surface. The CFRP was applied as done in situ (i.e. working under the beam). It is shown that the effect of CFRP strengthening was considerable. An analytical model has been used to simulate the load-deflection behavior as well as the failure mode of the pre-cracked RC specimens. Different failure mechanisms from ductile to brittle were simulated and verified by adopting the mechanical properties of the constituent materials. A description of two significant field applications concludes the paper by showing the feasibility of using the CFRP strengthening method.

Introduction

The need for strengthening (or stiffening) reinforced concrete (RC) and prestressed concrete (PC) structures is becoming more apparent, particularly when there is an increase in load requirements, a change in use, a degradation problem, or some design/construction defects. Potential solutions range from replacement of the structure to strengthening with composites such as fiber reinforced plastic (FRP) materials (ACI Committee 440 1996; El-Badry 1996). Among several possibilities, FRP composites can be used for this purpose in the form flexible carbon FRP (CFRP) sheets to be externally bonded to the concrete surfaces. The flexible sheets considered in this paper are made of unidirectional dry or pre-impregnated (pre-preg) fibers and are installed with a technique known as manual lay-up.

[1] Professor, Department of Architectural Engineering, The Pennsylvania State University, University Park, PA 16802, USA
[2] Research Associate, Department of Architectural Engineering, The Pennsylvania State University, University Park, PA 16802, USA

The paper first presents the results of experimental and analytical work aimed at the determination of the performance of strengthened members as well as their modeling. The paper then concludes with the description of two significant field applications.

Experimental Work

An experimental study (Arduini and Nanni 1997) was conducted consisting of four-point bending tests as well as coupon tests to characterize material properties including the concrete-adhesive interface strength. Two types of unidirectional CFRP material systems were used with the number of plies varying from one to three. The direction of the fibers was in most cases arranged parallel to the axis of the beam (longitudinal direction or 0 degree) in order to act as flexural reinforcement. In one case, CFRP sheets were wrapped around three sides of the beam (0 and 90 degree) for shear reinforcement and anchorage of the longitudinal sheets. Two concrete cross sections were investigated to represent the cases of shallow and deep sections. Two types of concrete surface preparation were used in order to understand the influence of this parameter on the failure mode. Finally, two beams were strengthened while carrying an applied load representing the expected service load of the member. The latter of these beams was also subjected to external postensioning during strengthening with the intent of compensating for the deflection due to service load.

In this paper the only results reported are relative to three beams (i.e., M1, MM2 and MM5) of the medium-length series with details and dimensions as shown in Figure 1. Specimens were constructed at a precast plant using conventional fabrication, curing, and transportation techniques. Specimen M1 did not have external FRP reinforcement. The concrete surface for the specimens MM2 and MM5 (to be strengthened with CFRP) was prepared by sanding and sand blasting, respectively. Beam MM2 was pre-loaded (pre-cracked) prior to the application of the FRP sheet using the same configuration of the test to failure (Figure 1). The applied pre-load was equal to 30% of the nominal capacity of the member to simulate a reasonable service condition and to allow the formation of three to four cracks in the region of constant moment. Both MM2 and MM5 were externally strengthened with prepreg sheets having the following mechanical characteristics: tensile strength equal to 580 N/mm of ply width, and tensile modulus equal to 24 kN/mm of ply width. As a point of reference, the thickness of an installed ply (which includes fibers and adhesive) is in range of 1 to 3 mm.

In the case of MM2 a single ply was used (950 mm long and 150 mm wide) on the beam soffit. In the case of MM5 two plies were used (850 mm long and 150 mm wide) on the beam soffit plus one ply on the sides at each 0 and 90 degree orientations. Adhesion of the FRP sheets was conducted by working under the specimen as it would be done in the field for the repair of the soffit of a flexural element. After seven days of resin curing, the beams were tested to failure. Results of this tests are shown in Figure 2 (see curves M1, MM2, and MM5) together with analytical results (see curves M1-A, MM2-A, and MM5-A).

Analysis

An analytical model (Arduini and Nanni 1997) was developed and its predictions show good correlation with experimental results (Figure 2). Different failure mechanisms from ductile to brittle (like FRP rupture or concrete shear failure at the edge of the FRP reinforcement), could be simulated and verified. The model takes into account the influence of concrete confinement in the compression zone due to the presence of the stirrups, and the tensile softening properties of concrete. This allows a more accurate determination of the crack propagation and the failure mechanism of the flexural member. The constitutive laws for the constituent materials considered by the analytical model are as follows:

- Concrete in compression is non-linear according to the CEB-FIP Model Code 90.
- For concrete in tension, the same non-linear relationship has been used until the peak strain, followed by a linear descending branch.
- Steel is bilinear elasto-hardening
- FRP and adhesive are perfectly linear elastic

With respect to the unloading-reloading constitutive laws, it is assumed that FRP and adhesive are perfectly linear-elastic and no permanent deformation results from cycling. This behavior is also adopted for steel when the cycle peak strain is less than the strain at yielding. When the peak strain exceeds that of yielding, the unloading-reloading constitutive law remains linear elastic with the slope of the initial E, but with a permanent deformation. For concrete in both compression and tension, the unloading-reloading constitutive relationships are detailed elsewhere (Arduini and Nanni 1997).

Field Projects

Building Application

The postensioned PC slab of a parking garage in Atlanta, Georgia, was strengthened with gunite RC beams shortly after construction in order to correct a deficiency in the number of steel tendons along the East-West alignment of the building.   These beams were 75 mm deep, 1 m wide and reinforced with 6 Number 9 (28.5 mm diameter) bars.   The beams were 5.2 m long and ran along the column line, connecting the column capitals (3 by 3 m).   The integrity of the composite action between gunite beam and slab was to be based solely on the strength of the interfacial bond between the two.   Since delamination had occurred over time, such action was compromised and epoxy injection required. In order to find a permanent solution to the problem, it was suggested that the gunite beams be demolished and replaced with two double-ply strips of CFRP.

Figure 3 shows the application of the second ply for one of the strips.   The two CFRP strips were located at the side of the demolished gunite beam so that adhesion would take place on a relatively smoother concrete surface.

In order to evaluate the condition of the PC slab with and without the gunite beams and also after strengthening with CFRP, a number of rapid load tests were carried out.   In the test set-up, a concentrated force was applied to the slab column-strip by means of hydraulic jacks (Figure 4).   As seen in the photograph, the jacks are reacting against the floor above, which in turn is shored for additional safety (Figure 5).   This configuration may be defined as a "push-type" test where the dead weight of the two floors above provides the counterweight.   Deflection at several points (e.g., under the load, at the quarter-span sections, at the drop panel) was measured (Figure 6).   Following repeated loading-unloading cycles, it is possible to develop a hysteresis diagram for the slab in its three different conditions.   The level of the maximum load was calibrated based on preliminary calculations and the response of the structure during test.   The load test was repeated with the same modality after the execution of the CFRP strengthening work.   The level of maximum load was then increased to simulate service conditions.   By comparing the outcome of the various tests, it was shown that the CFRP repair was comparable in strength to the gunite beams. However, CFRP should provide a permanent solution.

Highway Application

Figure 7 shows the effect of a vehicular impact on the four girders of the bridge overpass on highway Appia at km 1+344, near Terracina, Rome (Nanni 1997).   This is a short bridge, 10.5 m in span, made of four prestressed concrete girders having cross sectional dimensions of 1.0 by 1.5 m.   The conventional reinforcement (prestressing tendons and reinforcing bars) is clearly visible in the photograph after the loose concrete was removed.

The concrete cross section was restored with no-shrink mortar and, after surface preparation, CFRP sheets were adhered as shown in Figure 8. The objective of the CFRP strengthening was to make up for the loss of prestress. For each beam, three sheets, 0.33 m wide and 3.0 m long, were bonded to the soffit (0 deg. fiber direction), and four strips, 0.16 m wide and 3.0 m long, were wrapped around the three sides (90 deg. fiber direction). The total amount of CFRP material used was approximately 20 m^2. Figure 9 shows the completion of the job with the application of the finishing coat.

Conclusions

Based on the analytical and experimental results obtained in this study, some general conclusions can be drawn:

- The strengthening technology consisting of externally-bonded CFRP sheets is easy to perform and results in significant improvements in ultimate load capacity and, to a lesser extent, in flexural stiffness.
- An area in need of attention and, possibly, improvement is that of concrete surface preparation. It is necessary to avoid or at least limit the extent of FRP peeling in order to improve the effectiveness of the strengthening method and the ductility of the load-deflection response.
- It is possible to simulate and predict experimental load-deflection behavior, strain distribution, and the failure mode of FRP strengthened beams, including the effects of pre-cracking and unloading-reloading cycles (Arduini and Nanni 1997).

Two recently completed strengthening projects were presented to demonstrate that CFRP is becoming an acceptable rehabilitation method for buildings and infrastructures. As owners, designers, and contractors become familiar with this technology, it is envisioned that its number of applications will rapidly increase. The presented projects show the adaptability of CFRP technology to different situations: correction of design/construction errors and loss of integrity due to vehicular collision. As the technology matures, a field of application that is equally viable and important is that of damage prevention. In several countries interested in the use of FRP composites for construction, the lack of design and construction specifications will soon be overcome with the preparation of new codes and standards. In the interim, the practice of specific and well-thought load tests can be a powerful tool for the assessment of the damage and the evaluation of the rehabilitation work.

Appendix I: References

ACI Committee 440, 1996, "State-of-the-Art Report on FRP for Concrete Structures," ACI440R-96, Manual of Concrete Practice, ACI, Farmington Hills, MI, 68 pp.

El-Badry, M. (editor), August 1996, "Advanced Composite Materials in Bridges and Structures," Proceedings ACMBS-II, Montreal, Canada, pp. 1027.

Arduini, M. and A. Nanni, 1997,"Behavior of Pre-Cracked RC Beams Strengthened with Carbon FRP Sheets," ASCE-Journal of Composites for Construction (in print).

Nanni, A., 1997, "Carbon FRP Strengthening: New Technology Becomes Mainstream," Concrete International: Design and Construction, (in print).

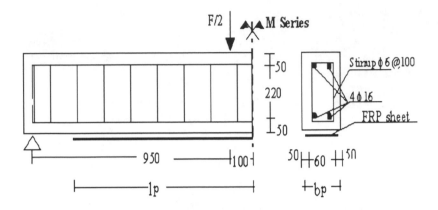

Figure 1: Beam Geometry and Set-up

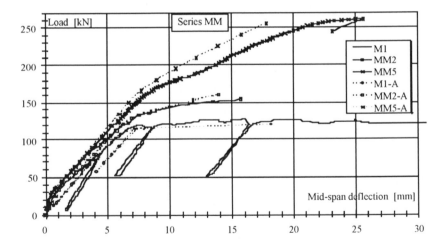

Figure 2: Experimental and Analytical Load-Deflection Curves

Figure 3: Application of Second CFRP Ply

Figure 4: Load Applied by Hydraulic Jacks

Figure 5: Shoring of the Floor Above Tested Slab

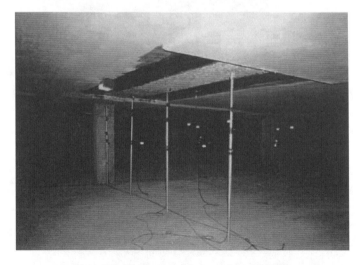

Figure 6: Deflection Measured by LVDT's

Figure 7: Girder Damage Due to Vehicular Impact

Figure 8: Pattern of CFRP Strips

Figure 9: Application of the Finishing Coat

Use of Inorganic Polymer-Fiber Composites
for Repair and Rehabilitation
of Infrastructures

by P. Balaguru[1] and Stephen Kurtz[2]

Abstract

This paper summarizes the properties of an inorganic polymer-fiber composite and the results of an experimental investigation of the behavior of reinforced concrete beams strengthened with carbon fiber fabrics and geopolymer. The primary objective of the investigation was to determine whether geopolymer can be used instead of organic polymers for fastening the carbon fabrics to concrete. Four reinforced concrete beams that were similar to the ones reinforced with carbon fabrics and organic adhesives were tested. The beams had 0, 2, 3 and 5 layers of unidirectional carbon fabrics attached at the tension face of the beams.

The results indicate that geopolymer provides excellent adhesion both to concrete surface and in the interlaminar planes of fabrics. All three beams failed by tearing of fabrics. This is very significant because very few researchers report failure of beams with tearing of fabrics. The most common failure pattern reported in the literature is the failure by delamination of fabrics near the interface of concrete and the composite or failure of concrete between reinforcing bars and composite. Hence it can be stated that geopolymer provides as good or better adhesion in comparison with organic polymers. In addition, geopolymer is fire resistant, does not degrade under UV light, and is chemically compatible with concrete. Therefore, the product can be successfully developed for use in the repair and retrofitting of concrete structures.

Introduction

It is well known that the national infrastructure is in need of major repairs and rehabilitation. A number of repair and strengthening techniques are being promoted. Strengthening of reinforced concrete structures with externally bonded steel plates is one of the techniques developed in the 1960's. Recently, high strength carbon, glass, and aramid composite plates are being promoted as a better alternative to steel plates [ACI, 1996]. The major advantages in using the composite

----

[1]Professor, [2]Research Assistant, Department of Civil and Environmental Engineering, Rutgers University, P. O. Box 909, Piscataway, NJ 08855-0909.

plates are: lightweight, corrosion resistance, and ease of application. The light-weight is a major advantage during construction because heavy equipment is not needed. The composites can also be applied layer by layer resulting in almost a homogeneous final structure.

The major disadvantage of composites is their lack of fire resistance and degradation under UV light leading to long-term durability problems. The carbon and glass fabrics can withstand normal fire exposure and are durable under UV light. But the weak link is the organic polymers that are used to attach these fabrics to concrete. Hence, an investigation was undertaken to evaluate the use of an inorganic polymer which was developed recently. This inorganic polymer, known as geopolymer, is an alumino silicate which can sustain up to $1000°C$ ($\approx 2000°F$). The polymer is durable and does not degrade under UV light. In addition the matrix is non-toxic and water based. Hence, cleaning and disposal of excess materials is not a problem. The matrix also provides a hard surface that cannot be easily vandalized. This aspect becomes important for structures located in urban areas.

The investigation under progress at Rutgers deals with mechanical properties of carbon and glass composites, and their use for aero structures and infrastructures (Foden, 1996). In the infrastructures area, the primary focus is strengthening of structural elements. The composite is also being evaluated for coating to improve durability of the existing structures.

The results obtained so far indicate that the inorganic polymer has excellent potential for use in infrastructures.

The primary conclusions obtained are as follows:

•   The system is very easy to work with and all the techniques used for organic polymers can be used for current matrix.

•   The matrix is compatible with carbon and glass fabrics. The carbon composite can sustain about 650 MPa, 550 MPa, and 30 MPa in tension flexure and shear.

•   The matrix adheres well to wood, concrete, and steel. The shear strength obtained using glued steel plates was 15 MPa.

•   The fatigue performance of carbon composite is comparable to the fatigue performance of organic polymer-carbon composite.

The results presented in this paper focuses on the behavior of strengthened reinforced concrete beams.

Research Program

A number of investigators have evaluated beams strengthened with carbon fibers and organic polymers (M'Ba Zaa, 1996, Nakamura, 1996). The current research program was designed to simulate the research conducted at the Universite de Sherbrooke [M'Ba Zaa, 1996]. This strategy was used to reduce the number of beams to be tested for comparing organic and geopolymers. Four singly reinforced concrete beams that were similar to beams used at Sherbrooke were cast and cured for 28 days. Then three of the beams were strengthened using carbon fabrics and geopolymer. All the four beams were tested as simply supported beams under four point loading. The details of the beams and experimental procedures are presented in the following sections.

Details of the Beams

Four reinforced concrete beams that were 3200 mm long, 200 mm wide and 300 mm deep were constructed. These beams were tested over a simply supported span of 3000 mm. The reinforcement details of the beams are shown in Fig 1. The tension reinforcement consisted of 2-# 4 (13 mm diameter) bars. The tension reinforcement was kept to a minimum, in order to avoid the shear failure of strengthened beams. The compressive strength of concrete was about 47 MPa. The control cylinders made during the fabrication of all four beams provided consistent compressive strengths.

Strengthening of Beams

Three beams were strengthened using 2, 3, and 5 layers of unidirectional carbon fabric. The fabric made of T300 carbon fibers had a density of 5 oz/yd^2. After curing, the bottom surface of the beams were roughened, first by dry grinding followed by sand blasting. These operations removed the weak mortar layer, exposing some aggregates.

The rough surface was primed with a mixture of geopolymer to avoid the loss of geopolymer from fabrics to voids in concrete. The fabrics themselves were pregged using hand pre-pregging and placed at the bottom surface of the beam. The beam with two layers was allowed to dry for 24 hours and heated to 80°C to cure the geopolymer. For beams with 3 and 5 layers, after placing the fabrics, they were covered with bleeding cloth and a vacuum of about 700 mm of mercury was applied for better adhesion. These beams were also heated to 80°C to facilitate curing. A typical strengthened beam is shown in Fig 2.

Instrumentation and Test Set-up

The beams were instrumented to measure strains in concrete, tension steel, and the composite; and the deflections. The strain values in the composite can be considered only as average values because the gages were glued to both the fibers and the matrix. The beams were simply supported over a span of 3000 mm and two concentrated loads were applied at 1000 mm from the supports. The loads were measured using MTS data logging systems. The beam set-up, ready for testing is shown in Fig 3.

The loads were applied in 4.4 or 2.2 kN increments. For each load-increment strains, deflections and crack pattern were recorded.

Results and Discussion

A summary of results is presented in Table 1, which shows loads corresponding to yield and final failure, and mid-span deflections at failure.

Mode of failure

As mentioned earlier, all the strengthened beams failed by rupturing of the composite. This shows that geopolymer provides effective adhesion even when five layers of fabric were used. In practice, the number of fabric layers have to be limited to 3 or 4 for economical reasons. Hence, if the repair system is properly carried out, failure by delamination of composite can essentially be eliminated.

Since the beams were purposely underreinforced with sufficient shear reinforcement, shear failure did not occur even when the moment capacity was increased by 50 percent over the control beam. As the number of layers increased, the length of composite that ruptures also increased.

Load-deflection behavior and crack patterns

The load-deflection curves have three segments representing; pre-cracking (of concrete), post cracking, and yielding (of steel) stages. The post yielding part of the curve for strengthened beams became shorter.

As expected, the stiffness of the beam increased with the number of layers of fabric as indicated by the decrease in deflection, shown in Fig 4. The depth of neutral axis seem to increase with the number of layers. This should be expected because increased tension force for a given curvature requires increased compression force. Since the strength of concrete is the same, the increased compression force capacity has to come from increased compression force provided by a larger depth of neutral axis.

The crack patterns of strengthened beams are different from the control beam. Strengthened beams had more cracks and were more closely spaced. As the number of layers increased, the length of the beam over which extensive cracking occurred also increased. Maximum crack widths were smaller for the strengthened beams. Typical crack patterns are shown in Figs 5, 6, 7 and 8.

Comparison of organic and geopolymer

As mentioned earlier, the beams were designed so as to allow direction comparison of results obtained by Labossiere et al. at the Universite de Sherbrooke. Their control beam had a capacity of 63.8 kN, and their strengthened beam had a capacity of 99.8 kN. Hence the strengthening provided an increase of about 50 percent. They used 3 layers of Tonen Unidirectional fabric. The amount of reinforcement in 3 layers of Tonen fabric is slightly higher than 5 layers of fabric used in the current study. The beam with 5 layers also sustained 50 percent more load than the control beam, Table 1. The deflections were smaller for beams with geopolymer. The beam with 5 layers failed at a deflection of 23 mm as compared to 28 mm deflection reported in the Sherbrooke study.

The primary difference between the organic polymer and the geopolymer is the failure pattern. In the Sherbrooke study, the composite peeled off, whereas the composite ruptured in the current study. Delamination failure not only underutilizes the composite strength, but is also extremely brittle. This type of failure must be avoided at all costs, in order to provide warning of the impending failure.

The deflections and crack patterns of beams with organic and geopolymers are comparable. The composite in the current study recorded larger strains than the strains reported in the Sherbrooke study.

In summary, it can be stated that geopolymer provides a better structural performance than the organic polymer.

Conclusions

Based on the experimental results obtained in the current study, and the results reported by other researchers, the following conclusions can be drawn.

• Geopolymer can be successfully used to bond carbon fabrics to reinforced concrete beams.

• With proper design and construction process, failure by delamination of composite can be eliminated.

• The performance of geopolymer is better than organic polymer in terms of adhesion. In addition, geopolymer is fire resistant, durable under UV light and does not involve any toxic substances. Geopolymer is water based and no special protective equipment other than gloves is needed. Excess material can be discarded as ordinary waste. This aspect is very important during the construction phase.

References

1.   ACI Committee 440, State-of-the-Art Report on Fiber Reinforced Plastic Reinforcement for Concrete Structures, American Concrete Institute, Detroit, MI, 1996, 68 pages.

2.   Foden, A., Balaguru, P., and Lyon, R., "Mechanical Properties of Carbon Composites   Made Using an Inorganic Polymer," ANTEC, 1996, pp. 3013-3018.

3.   M'Ba Zaa, I., Missihoun, M., and Labossiere, "Strengthening of Reinforced Concrete Beams with CFRP Sheets," Fiber Composites in Infrastructure, 1996, pp. 746-759.

4.   Nakamura, M., Sakai, H., Yagi, K., and Tanaka, T., "Experimental Studies on the Flexural Reinforcing Effect of Carbon Fiber Sheet Bonded to Reinforced Concrete   Beam," Fiber Composites in Infrastructure, 1996, pp. 760-773.

Table 1.   Summary of Test Results

Beam Design.	Load at Yielding of Steel, kN	Failure Load, kN Exp.	Deflection at at Failure, mm	Mode of Failure
Control	55.6	71.2	89	Yielding of steel
With 2 layers	62.3	80.5	19	Rupture of Composite
With 3 layers	70.3	91.2	23	Rupture of Composite
With 5 layers	73.4	109.9	23	Rupture of Composite

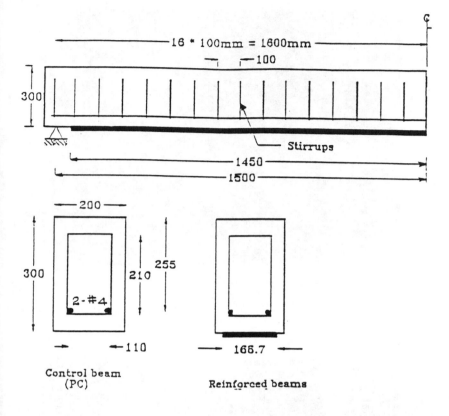

Figure 1 Beam Cross Sections.

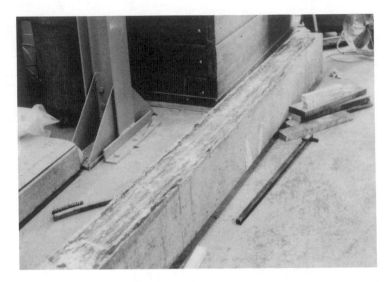

Figure 2a. Beam After Application of Primer

Figure 2b. Beam After Strengthening

Figure 3 Beam Number Two Prior to Start of Testing

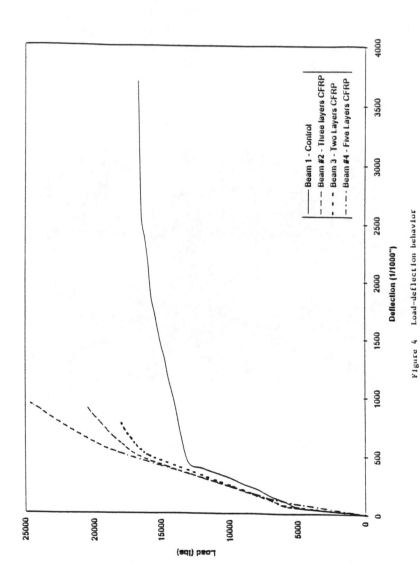

Figure 4   Load-deflection behavior

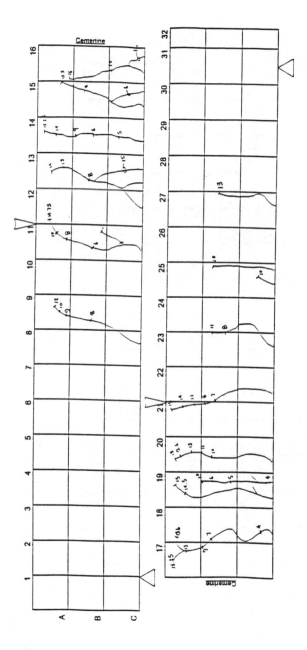

Figure 5   Crack Pattern, control beam

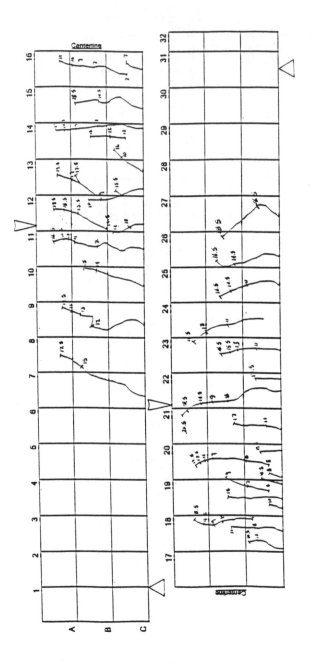

Figure 6   Crack Pattern, beam with two layers of fabric

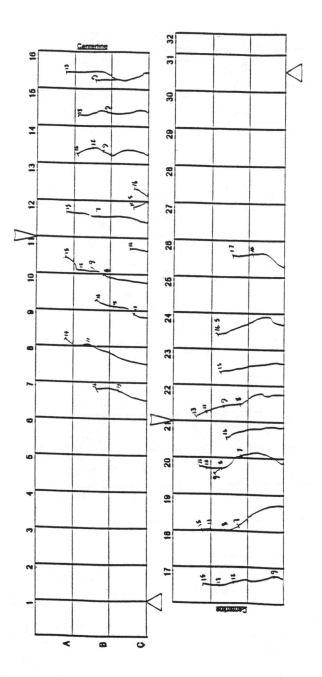

Figure 7   Crack Pattern, beam with three layers of fabric

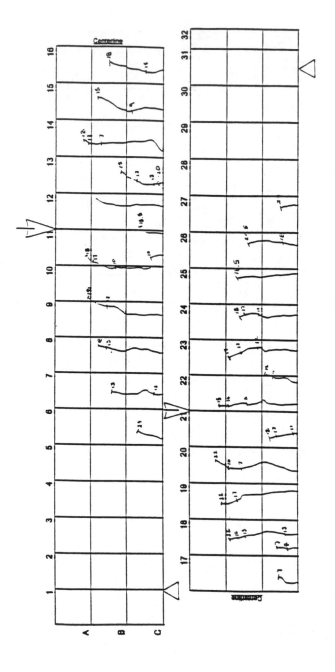

Figure 8   Crack Pattern, beam with five layers of fabric

A Case of Deterioration of Concrete Blocks Because of a Zeolite in the Aggregate

Silvio Delvasto[1], Ruby Mejia de Gutierrez[1], Carmen Elisa Guerrero[2], and Herman Klahr[3].

Abstract

This paper is the result of the investigation carried out at the Departamento de Materiales de Ingenieria of the Universidad del Valle to identify the cause of distress in a lot of masonry blocks produced by a local manufacturer.

Two phases were observed in the mass of the blocks. One, grayish, and the other, a white fluffy phase, distributed through the volume of the blocks. Soundness and Potential Reactivity tests were conducted to evaluate the coarse aggregate. The mixing water was analyzed to determine its acidity, and the content of chlorides and sulfides. The chemical and mineralogical composition (by x - rays diffraction) were determined for the fluffy phase and for the white part of the aggregate.

It was concluded that the presence of laumontite, a white zeolite, in the aggregate was the main reason for the concrete blocks crumbling. This mineral eventually loses water, becomes opaque, and disintegrates gradually on exposure to air. This finding demonstrates that another cause of disruption of concrete arises from the interaction between the mineral phases of the aggregates and their surroundings, one of them but not the unique is the alkali aggregate reaction

[1] Titular Professor, Departamento de Materiales de Ingeniería, Universidad del Valle, A.A.25360. Cali, Colombia
[2] Research Technical Assistant, Departamento de Materiales de Ingeniería, Universidad del Valle, A.A.25360. Cali, Colombia
[3] Civil Engineer, Manager of Blokes Ltda, Cali, Colombia

Introduction

Unlike in some other countries, aggregate reaction has not received a great deal of attention in Colombia. Only alkali reaction has been mentioned occasionally. At the present time, as far of the authors know, no case of distress of structures during service life caused by reactivity of aggregates have been reported. However, it has been found some problems of deleterious expansions in precast concretes, more of them related with the utilization of coal slags as lightweight aggregate and in other cases carbonation due to high porosity.

It is recognized that several types of aggregate reactions could happen. However, most of the authors related them as aggregate reactions involving alkalis (Campbell-1983). In general, these reactions occur anywhere from a few months to many years of concrete service (Mullick-1988) (Danay-1994). As a result of the chemical attack, an alkaline silica gel is formed. Swelling of this gel produces dimensional changes and consequently expansive pressures within the concrete. If the gel is of sufficient rigidity it generates expansive deterioration when the gel cannot be accommodate in the surrounding porous space of the concrete mass.

In a local factory of building materials, signs of disruption of a lot of masonry concrete blocks were observed accompanied by white popouts after several weeks of curing. An engineer, manager of the precasting plant, went to the Universidad del Valle seeking for answers to that problem. This paper is the result of the investigation carried out at the Departamento de Materiales de Ingenieria of the Universidad del Valle to identify the cause of distress in the case of these masonry blocks.

Experimental Work

After visiting the precast plant, selected samples of the reacted blocks and their raw materials were taken. External manifestations of distress included occurrence of white deposits distributed through the mass of the block. The concrete of the block easily disintegrated gradually into small fragments by applying hand pressure. A typical distressed block and crumbs coming from another are shown in Figure 1. Two phases were observed in samples of the concrete blocks by visual examination of the broken surface. One, grayish, and the other, a white fluffy phase, distributed through the volume of the blocks. After samples of the reaction product, a white fluffy - powdered gel, were removed carefully with a blade, only remains the gray hard phase of the aggregate. The popout reaction phase and the blocks raw materials were studied after discarding the mixed ordinary portland cement to be the reason of the deterioration.

Figure 1. Mansory Blocks Sample

Soundness and Potential Reactivity tests were done to the aggregate. Also, in order to determine the effect of mixing water, its acidity, and the content of chlorides and sulfides were measured. The chemical and mineralogical properties of the fluffy phase and of the white part, around the gray diabasic aggregates, were determined by standard methods and by X-ray diffraction (XRD).

Results and Discussion

          ASTM C150 and DIN 1164 accept an upper limit
of 0.6 % Na ᵢ O equivalent for cement used in
combination with reactive aggregates. Although,
expansions have been observed with an Na ᵢ O equivalent
as low as 0.4 %. Chemical composition of the fluffy
powder and a white part of the diabasic aggregate
proportioned in the original mix of the blocks are
presented in Table 1. From that can be established
that the total amount of alkalis as Na_2O equivalent,
0.20 % in the fluffy part, is small enough to have been
caused by a reaction between aggregate particles and
the alkalis. It is quoted that there is a probability
that this amount of alkalis could produce an expansive
alkali - carbonate reaction because it seems that the
alkali ions play a role as catalysts in the alkali -
carbonate reaction (Mingshu-1994). However, it is not
observed modification of the calcite crystalline
structure after the disintegration of the masonry
blocks as is proved by x - rays diffraction.

          In addition, as can be seen in Table 2, the
soluble silica, S , 9.16 millimoles / liter is lower
than the reduction of alkalinity, R , 173.61 millimoles
/ liter. In accord with the Standard ICONTEC 175, If
R is below 70 millimoles / liter and S < R the
aggregate does not present reactivity with the alkalis.
In addition, the mineralogical analysis of the fluffy
product does not show significantly compounds with
alkalis. Thus, problems of reactivity between aggregate
and alkalis are discarded.

          The soundness testing shown in Table 3
reported a range of loss of weight between 8.0 % and
10.2 %, below but near the limit value established by
ICONTEC 174 of 12.0 % of mass weight at five cycles of
being attacked by anhydrous sodium sulfate. The
results of the aggregate soundness test match probably
better with the mineralogical composition according to
the high losses of weight obtained. Additionally, the
unsoundness of rocks is related to pore size
distribution, which affects the aggregate strength
(Chang-1996).

Table 1 Chemical Composition of Block Fluffy Gel and Diabase White Part

% WT	Fluffy Gel	White Part of Aggregate
Si O_2	51.89	50.61
Al $_2O_3$	20.12	15.97
Fe $_2O_3$	0.36	7.04
Ca O	12.92	16.07
Mg O	0.00	1.03
Na $_2$O	0.14	-
K $_2$O	0.094	-
Na $_2$O equivalent (% Na $_2$O + 0.658 % K $_2$O	0.20	-
LOI	21.83	9.22

Table 2 Potential Reactivity of Diabasic Aggregate

Soluble silica (S_c) , millimoles / liter	9.16
Reduction of alkalinity, millimoles / liter	173.61

Table 3 Soundness Testing of aggregate

Fraction No.	ASTM Sieve No.	% Loss of material
1	4	10.2
2	8	8.0
3	16	8.7
4	30	9.5

The mixing water had a pH of 6.56 which is normal for a water to produce any effect on the portland cement behavior. The chloride ion concentration, 0.09 g/l is very low compared with the standards upper limit of 6 g/l. Similar situation happened to the sulfate ion concentration, 76.1 mg/l, compared to the standards limits, 1000 mg/l. Thus, the mixing water was appropriated for the concrete preparation.

The determination of mineral phases in the diabasic aggregate by x - rays using a RIGAKU Rint 2100 diffractometer shows, in order of quantity: laumontite, quartz, calcite, albite, augite, chlorite, prehnite, and almandine (traces). The fluffy product of reaction was also studied by x - rays diffraction. This fraction also presents abundantly laumontite and secondarily, calcite. The chemical composition of both matched narrowly. Indeed, from the obtained results, the fluffy phase was derived from part of the white backing rock of the original aggregate.

The laumontite, $Si_4O_{12}Al_2Ca.4 H_2O$, is a white zeolite. The stoichiometric chemical composition of the laumontite is given in Table 4. This composition agreeds with that of the fluffy material which was presented in Table 1, particularly in the content of silica, alumina, and calcium dioxide. The difference in Loss of Ignition is explained by the high content of carbonates coming from the calcite.

Table 4 Stoichiometric Chemical Composition of
Laumontite

Si O_2	51.09
Al $_2O_3$	21.68
Ca O	11.92
H$_2$ O	15.30

The laumontite, also called laumonite or lomontite, presents a holohedral monoclinic crystalline structure having the hightest symmetry in each crystal class. Generally, it is found in vesicles of eruptive rocks, specially basics rocks as diabases.It must be noted that the laumontite also has been found in a diabase of New Jersey (Klockmann-1961). The laumontite, because of its zeolitic characteristics, ion - exchanges large cations and loosely held water molecules allowing reversible dehydration. Because of this, the laumontite eventually loses water, becomes opaque, and disintegrates gradually on exposure to air.

The reversible dehydration of the laumontite is expansive and becomes puffed up exerting pressure which can crack the concrete. When expansive tensile strains over the tensile strain capacity of concrete are superimposed, the long term effects of these internal stress can be easily appreciated, not specifically because of visible evidence of cracking, but for external manifestations of spread white crumbs.

Sarkar and Aïtcin emphasize after carrying out characterization of twelve different coarse aggregates, that the petrological evaluation of other possible detrimental features is of significant importance in determining the aggregate's suitability for very high - strength applications (Sarkar-1990). Past characterization of aggregates has been limited to maximum size and grading, the experience shown in this paper confirm the necessity to pay attention to the petrological and mineralogical characterization of aggregates in order to prevent catastrophic disruptions of the concrete elements.

Concluding Remarks

It is concluded that the presence of laumontite, a white zeolite mineral, in the aggregate was the main reason for the concrete blocks crumbling. This finding demonstrates that aggregates could be reactive to some degree. It is necessary to learn that another cause of disruption of concrete arises from the interaction between the mineral phases of the aggregates and their surroundings, one of them but not the unique is the alkali aggregate reaction, as it is proved in this paper. The aggregate investigated here must be classified as potentially reactive. The use of such aggregate have to be avoid by procedures as selective quarrying.

References

1. Campbell - Allen, D. and Roper, H. "Selection of materials to improve performance of concrete in service". Handbook of Structural Concrete. McGraw - Hill Book Company. Great Britain, 1983, pp. 4-1 to 4-42.

2. Chang, T., and Su, N., "Estimation of coarse aggregate strength in high-strength concrete". ACI Materials Journal, V. 93, N° 1, January - February 1996, pp. 3-9.

3. Danay, A. "Structural mechanics methodology in diagnosing and assessing long - term effects of alkali - aggregate reactivity in reinforced concrete

structures". ACI Materials Journal, V.91, No. 1, January - February 1994, pp. 54 - 62.

4. Klockmann, F. and Ramdohr, P. Tratado de Mineralogía. Editorial Gustavo Gili, S. A. Barcelona, 1961, pp. 667.

5. Mingshu, T., Min, D., Xianghui, L. and Sufen, H. "Studies on alkali - carbonate reaction". ACI Materials Journal, V. 91, No.1, January - February 1994, pp. 26 - 29.

6. Mullick, A. K. "Distress in a concrete gravity dam due to alkali silica reation". The International Journal of Cement Composites and Lightweight Concrete, Volume 10, Number 4, November 1988, pp. 225 - 32.

7. Sarkar, S. L. and Aïtcin, P. C. "The importance of Petrological, petrographical and mineralogical characteristics of aggregates in very high strength concrete". Petrography Applied to Concrete and Concrete Aggregates, ASTM STP 1061, Bernard Erlin and David Stark, Editors, American Society for Testing and Materials, Philadelphia, 1990.

Application of Ambient Vibration Measurements for Repair of the Chaco-Corrientes Cable-Stayed Bridge in Argentina

Carlos A. Prato[1], Marcelo A. Ceballos[2],
Pedro J.F. Huerta[3] and Carlos F. Gerbaudo[3]

Abstract

The Chaco-Corrientes cable-stayed bridge, completed in 1973, is currently undergoing a repair program consisting on replacement of all cable stays and restitution of the longitudinal profile affected by significant permanent vertical deflections primarily due to creep of its concrete deck.

To define the degree of profile restitution compatible with acceptable safety margins of its main longitudinal box girders and cables, both the current stiffness of the main structural components and their state of stress had to be determined.

The current stiffness of the concrete deck was determined by means of ambient vibration recordings of the bridge deck under traffic and wind forces. Measured natural frequencies are used to calibrate the effective stiffness of a numerical model, which in turn is used to determine the additional forces in cables and girders required for restitution of the longitudinal profile.

Introduction

The Chaco-Corrientes cable-stayed bridge is a vital link of a major roadway connection across the Paraná River. Inaugurated in 1973, its structure consists of two main longitudinal box girders of prestressed concrete segmental construction. Each pylon is a cast-in-place reinforced concrete space frame

---

[1]Professor, National University of Córdoba, Argentina
[2]Research Assistant, National University of Córdoba, Argentina
[3]SETEC SRL, Consulting Engineers, Independencia 533, CP 5000, Córdoba, Argentina

rigidly connected to the box girders at deck level, and supported by thirty two 1.80 m diameter concrete piles. A general view is given in the photo of Figure 1.

Figure 1. Chaco-Corrientes Bridge

The static scheme of the bridge consists of two independent half bridges as shown in Figure 2, connected between them and to the access viaducts by simply supported spans of 20 m. Each half bridge is approximately symmetric about the vertical plane containing the axis of the pylon. The central span measured at centerlines of pylons is 245 m and the lateral spans are of 175 m.

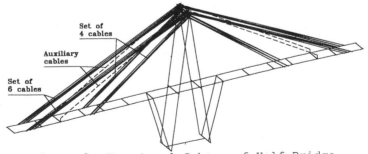

Figure 2. Structural Scheme of Half Bridge

The pylons are cast-in-place reinforced concrete space frames; the main box girders within the pylons are also of cast in place reinforced concrete with rigid connections to the other members of the pylon. The box girders outside the pylons are made of precast prestressed concrete segments. Similarly, both the deck slab and transverse beams that connect the box girders are made of precast prestressed concrete elements.

The original cables are of locked-coil type with external layers of galvanized wires. The cables are organized in sets of 6 cables for the longer stays and of 4 cables for the shorter ones. All cables have a centering collar between 2 to 3 m from the ends to reduce bending at the sockets.

The main structural issues that have required maintenance work during 24 years of service are the following:

i) Replacement of the elastomeric supports of the simple spans connecting the two half bridges and to the access viaducts. The original supports crept out of position during the first 10 years of service due to a combination of delayed shortening of the box girders, and of vertical/transverse oscillations induced by traffic loads and wind. Replacement of these supports was done approximately 10 years ago.

ii) Permanent vertical deformations of the bridge deck caused primarily by creep of the concrete deck. The maximum vertical displacement at the ends of the half bridges are very similar at all four extreme sections and equal to approximately 28.5 cm. These deformations affect the flow of traffic over the bridge.

iii) Failure of several Z-shaped wires of the external layer of the locked-coil cables occured in 1986 when adjustments of cable lengths were to be introduced to correct the permanent vertical deformations. This type of damage was evidenced in only four cables (of a total of 80) of the groups of 6. After a series of controls by magnetic induction confirming that wire failures were basically circunscribed to six cables, replacement of all stay cables was initiated in August of 1996.

The new cables are made of thirty three strands (T15) of galvanized steel with individual sheaths filled with wax. They are cut at the site and installed strand by strand in a sequence designed to achieve uniform distribution of stress within a stay using the so called "isotension" procedure.

Ambient Vibration Measurements

The ambient vibration records, taken just before starting cable replacement, consist of a series of vertical acceleration time histories of three representative points at the curb of the roadway. These points, located above the longitudinal axis of the main girders and very close to the vertical plane containing these axes, are in correspondence with three characteristic cross sections:

Sensor 1: Cross section passing through the intersection of the longitudinal axis of the girder and the resultant force of the groups of 6 stays.

Sensor 2: Cross section passing through the intersection of the auxiliary stays (not yet installed when the records were made) and the longitudinal axis of the girder.

Sensor 3: Cross section passing through the intersection of the longitudinal axis of the girder and the resultant force of the groups of 4 stays.

These sets of 3 simultaneous records, made along both main girders in the two half bridges, were repeated several times at each location to provide a variety of forcing functions during intervals with appreciable oscillations induced by wind and traffic. The frequency content of these functions depends on vehicle characteristics and transported weight, velocity and distance between vehicles, rugosity of the pavement and expansion joints and the characteristics of the wind.

The acceleration sensors are inductive accelerometers (Hottinger Baldwin) linear in the frequency range of 0 to 100 Hz, with a drop of 3 db at 200 Hz. They were connected to a signal conditioner and amplifier prior to digitization and storage on hard disk. These accelerometers are capable of sensing free vibrations of the fundamental mode of the bridge at 0.50 Hz, well below the range of most piezoelectric accelerometers and displacement sensors in the market.

All recordings were made with a time interval of 0.002 s, with a total duration of 7 seconds in order to pick up at least three cycles of the fundamental frequency. A typical record is shown in Figure 3.

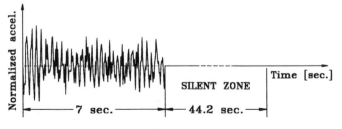

Figure 3. Typical Acceleration Record

Numerical Model

A numerical model of the bridge was developed
using prismatic bar elements as shown in Figure 2. This
model was intended to serve several purposes, mainly
to: i) Analyze the forces and displacements associated
with temporary loading conditions that occur during
replacement of the cables; ii) Determine the maximum
deck forces associated with the geometrical corrections
of the longitudinal profile of the bridge, and iii)
Assist in the interpretation of the vibration
measurements and derive the effective modulus of
elasticity of the concrete members.

It should be noticed that even though the
properties of concrete members are symmetric about the
pylon axes, the vertical curve of the bridge introduces
sligth asymmetries in cable lengths and global
stiffness of the two ends of each half bridge.

The geometrical properties of the bridge were
obtained from design drawings. Nodal coordinates were
adopted according to the ideal or "exact" design
geometry, since actual deviations of elevation with
respect to the design do not lead to appreciable
differences of results of the model.

The actual mass of the bridge is another relevant
data set. Dimensional controls of the thickness of the
box girders and road surface indicate that the weight
of the deck is at present 15-17 % larger than
anticipated from design dimensions in general, plus
some 40 t of additional weight at each end of the half
bridges introduced during the previous stage of repairs
of the simple spans (1986-87).

Recognizing that some of the additional weight
modifies the theoretical cross sectional properties,
the ideal member properties were used in the model,
except for the corrections of the masses that were
introduced explicitly.

The modal shapes and frequencies of the bridge are
given in Figure 4 for the lowest 30 modes. The modulus
of elasticity used for analysis is that resulting from
an optimization procedure discussed later.

Results of this analysis are helpful to explain some of the features of the observed behavior of the bridge. Among them:

a) The mode with the lowest frequency corresponds basically to a rotation of the deck about the axis of the pylon. This mode is not easily excited by traffic loads, but it can be excited by asymmetric lateral wind pressures and cause appreciable horizontal displacements. These lateral vibrations induce concentrated deformations on the supports of the simple spans, that added to the longitudinal shortening of the main girders were responsible for failure of the elastomeric supports.

b) The frequencies of modes involving significant torsional deformations are sensitive to the stiffness properties of members joining the two main box girders. This effect is consistent with the conclusions derived from this model when used to assess the static deformations associated with asymmetric loading conditions originating from stressing of the auxiliary cables and/or removal of existing cables. For this reason, and recongnizing the inherent limitation of the model to accurately represent the transverse diaphragms and deck slab, these modes were not used for determination of the stiffness of the longitudinal box girders.

c) The natural modes involving almost pure vertical bending deflections appear in pairs of symmetric and antisymmetric groups. This is the case of modes 2 (asym.) and 4 (sym.), modes 9 (sym.) and 10 (asym.), modes 14 (sym.) and 15 (asym.)and others. These modes, easily excited by the traffic loads and prominent in the vertical acceleration measurements, are the only ones used to determine the effective tiffness of the bridge.

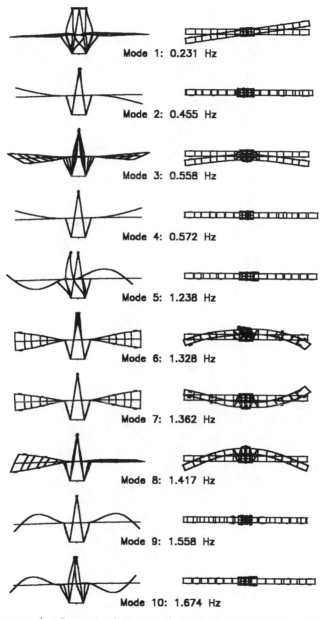

Mode 1: 0.231 Hz

Mode 2: 0.455 Hz

Mode 3: 0.558 Hz

Mode 4: 0.572 Hz

Mode 5: 1.238 Hz

Mode 6: 1.328 Hz

Mode 7: 1.362 Hz

Mode 8: 1.417 Hz

Mode 9: 1.558 Hz

Mode 10: 1.674 Hz

Figure 4. Computed Natural Modes and Frequencies

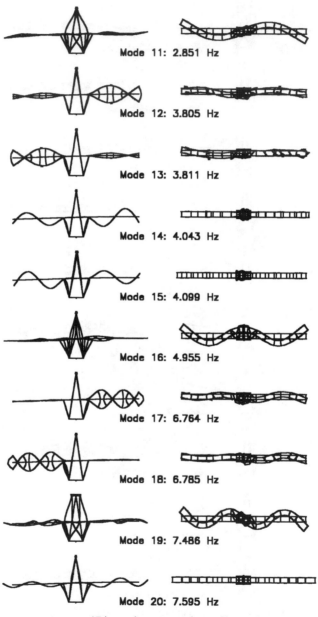

Mode 11: 2.851 Hz

Mode 12: 3.805 Hz

Mode 13: 3.811 Hz

Mode 14: 4.043 Hz

Mode 15: 4.099 Hz

Mode 16: 4.955 Hz

Mode 17: 6.764 Hz

Mode 18: 6.785 Hz

Mode 19: 7.486 Hz

Mode 20: 7.595 Hz

(Fig. 4 - continued)

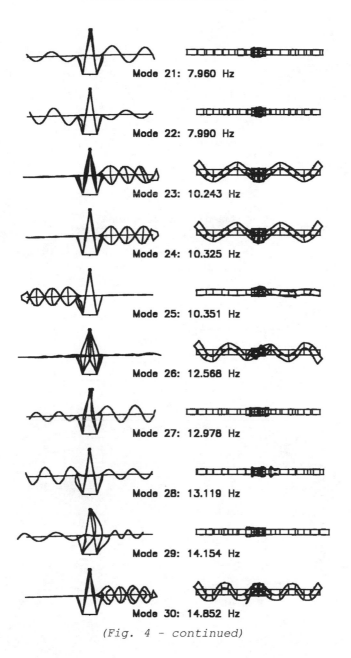

Mode 21: 7.960 Hz

Mode 22: 7.990 Hz

Mode 23: 10.243 Hz

Mode 24: 10.325 Hz

Mode 25: 10.351 Hz

Mode 26: 12.568 Hz

Mode 27: 12.978 Hz

Mode 28: 13.119 Hz

Mode 29: 14.154 Hz

Mode 30: 14.852 Hz

(Fig. 4 - continued)

Interpretation of Experimental Results

    Identification of the natural frequencies from the
recorded accelerograms was carried out with the
spectral density function of the records. To eliminate
peaks of these individual records due to the
characteristics of the forcing function of each record,
and to retain only the common ones associated with the
natural frequencies, the average power spectral density
function was calculated for each sensor normalizing the
amplitude of all records to the same root mean square
value (Felber et al.[1], Cantieni [2]). The average
normalized power spectral density function (ANPSD) for
each typical sensor location is given in Figure 5.

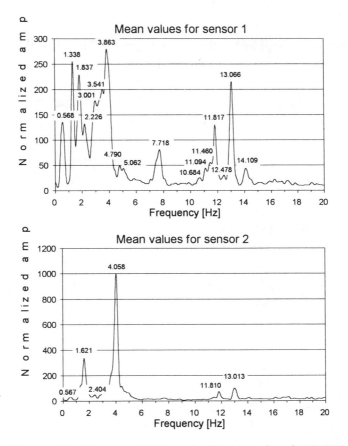

Figure 5. Measured Natural Frequencies by ANPSD

The natural frequencies are also identified by means of the Phase Dispersion of the records as defined by Ceballos et al. [3], which itself is a normalized function. The results of the analysis are given in Figure 6. The frequencies identified by these procedures from sensor 2 are given in Table 1. The two procedures yield values of the natural frequencies in close agreement.

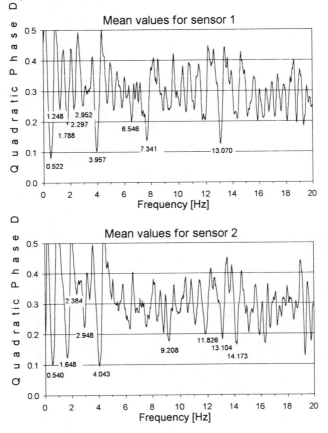

Figure 6. Measured Natural Frequencies by Phase Dispersion

Table 2 contains the natural frequencies of the numerical model computed for several values of the modulus of elasticity of concrete at regular intervals of 500 MPa. Table 3 contains the averaged frequencies

of the group of modes   (a),   (b)   and   (c)   that   are
used   for

Frequency of picks	ANPSD	Phase Dispersion
1	0.567	0.540
2	1.621	1.648
3	2.404	2.384
4	-	2.948
5	4.058	4.043
6	-	9.208
7	11.810	11.826
8	13.013	13.104
9	-	14.173

Table 1. Comparison of Natural Frequencies by ANPSD
and Phase Dispersion Procedures from Sensor 2.

Mode	Frequency [Hz]								
	Modulus of elasticity: .10^4 MPa								
	4.00	4.05	4.10	4.15	4.20	4.25	4.30	4.35	4.40
1	0.224	0.225	0.227	0.228	0.230	0.231	0.232	0.234	0.235
2	0.447	0.448	0.450	0.451	0.453	0.455	0.456	0.458	0.459
3	0.542	0.545	0.548	0.552	0.555	0.558	0.562	0.565	0.568
4	0.569	0.570	0.570	0.571	0.572	0.572	0.573	0.573	0.574
5	1.206	1.213	1.219	1.225	1.232	1.238	1.244	1.250	1.256
6	1.293	1.300	1.306	1.314	1.321	1.328	1.334	1.341	1.348
7	1.328	1.334	1.340	1.348	1.354	1.362	1.367	1.374	1.380
8	1.376	1.384	1.393	1.401	1.409	1.417	1.425	1.433	1.441
9 (a)	1.521	1.529	1.536	1.544	1.551	1.558	1.566	1.573	1.580
10 (a)	1.628	1.637	1.647	1.656	1.665	1.674	1.683	1.692	1.701
11	2.766	2.783	2.800	2.817	2.834	2.851	2.867	2.884	2.900
12	3.693	3.714	3.735	3.759	3.781	3.805	3.824	3.846	3.866
13	3.699	3.720	3.741	3.765	3.786	3.811	3.829	3.851	3.872
14 (b)	3.924	3.948	3.972	3.996	4.020	4.043	4.067	4.090	4.113
15 (b)	3.978	4.002	4.026	4.051	4.075	4.099	4.122	4.146	4.170
16	4.808	4.838	4.867	4.897	4.926	4.955	4.984	5.013	5.042
17	6.559	6.599	6.638	6.681	6.721	6.764	6.800	6.840	6.878
18	6.579	6.619	6.658	6.701	6.741	6.785	6.821	6.861	6.899
19	7.263	7.308	7.353	7.398	7.442	7.486	7.530	7.574	7.617
20	7.373	7.418	7.463	7.507	7.551	7.595	7.638	7.681	7.724
21	7.724	7.772	7.819	7.867	7.914	7.960	8.007	8.053	8.099
22	7.753	7.801	7.849	7.896	7.943	7.990	8.037	8.083	8.129
23	9.938	9.999	10.060	10.122	10.182	10.243	10.302	10.362	10.421
24	10.015	10.076	10.137	10.200	10.261	10.325	10.382	10.443	10.502
25	10.041	10.102	10.162	10.226	10.287	10.351	10.409	10.469	10.528
26	12.193	12.269	12.344	12.419	12.493	12.568	12.641	12.714	12.787
27 (c)	12.591	12.669	12.747	12.824	12.901	12.978	13.054	13.129	13.204
28 (c)	12.728	12.807	12.885	12.964	13.041	13.119	13.195	13.272	13.348
29	13.732	13.818	13.902	13.987	14.070	14.154	14.236	14.319	14.400
30	14.408	14.497	14.585	14.675	14.763	14.852	14.938	15.024	15.110

Table 2. Computed Natural Frequencies vs. Effective
Modulus of Elasticity of Concrete.

Mode	Selected control frequencies [Hz]								
	Modulus of elasticity: .10^4 MPa								
	4.00	4.05	4.10	4.15	4.20	4.25	4.30	4.35	4.40
a	1.575	1.583	1.591	1.600	1.608	1.616	1.624	1.633	1.641
b	3.951	3.975	3.999	4.023	4.047	4.071	4.095	4.118	4.141
c	12.659	12.738	12.816	12.894	12.971	13.048	13.125	13.201	13.276

Table 3. Selected Control Frequencies for Comparison
with Measured Ones.

correlation with the measured frequencies. Table 5
gives a comparison between computed and measured
natural frequencies and mode shapes at sensor locations
1 and 2. The measured mode shapes are defined as the
deformations evaluated at the natural frequencies
(ANPSD) and thus are really linear combinations of the
response modes. It is natural then that mode (b) is
found closer than modes (a) and (c) to the computed
modes due to its large participation in response to
ambient forces.

The effective modulus of elasticity of the
concrete deck was determined by minimizing the
difference between calculated and measured natural
frequencies of the 3 dominant modes of response at
sensor 2. The records of sensor 2 were preferred over
those of the other locations because bending modes are
more clearly picked up there.

The error to be minimized was defined as the sum
of the absolute values of the difference between
measured and computed frequency of modes (a), (b) and
(c), weighted by the square root of the amplitude of
the PSD function of the corresponding mode. This was
intended to assign more weight to those modes with
larger amplitudes in the records. Alternatively, a
variant involving the difference of frequency divided
by the frequency was also used to avoid a bias towards
the higher frequencies. The variation of the error
function with these criteria, denoted as I and II
respectively, is given in Table 4. The value of the
modulus of elasticity of concrete leading to a minimum
error was found to be 42500 MPa with both alternatives
of the error function.

It is emphasized that this value of the modulus of
elasticity of concrete should be regarded as an
effective modulus applicable for structural analyses

with the numerical model with all other parameters
fixed, rather than a measure of the stiffness of the
material. In fact,

Criterium	Best fit criteria								
	Modulus of elasticity: .10^4 MPa								
	4.00	4.05	4.10	4.15	4.20	4.25	4.30	4.35	4.40
I	0.130	0.101	0.073	0.045	0.017	0.014	0.039	0.066	0.094
II	0.027	0.021	0.016	0.010	0.004	0.003	0.007	0.012	0.018

Table 4. Determination of the Effective Modulus of
Elasticity of Concrete.

Mode	Sensor	Computed			Measured		
		Freq. [Hz]	Norm. amp.	Phase [°]	Freq. [Hz]	Norm. amp.	Phase [°]
a	1	1.616	0.201	0	1.621	0.679	0
	2		1.000	180		1.000	175
b	1	4.071	0.328	0	4.058	0.326	0
	2		1.000	0		1.000	0
c	1	13.048	0.931	0	13.013	1.204	0
	2		1.000	180		1.000	165

Table 5. Comparison of Control Modal Shapes at
Sensors 1 and 2.

the modulus of elasticity of concrete determined with
the ACI code for the strength of recently cored samples
is 38000 Mpa.

Effective Modulus from Static Measurements

      The efective modulus of concrete was also
determined by measuring vertical deflections through
levelling of the deck when a cable stay is removed. The
observed vertical deflection at the end sections is
approximately 0.07 m, and corresponds to a concentrated
force of approximately 290 t. This deflection is
sufficiently large to turn the error in levelling not
significant (+/- 0.002 m).

      The measured vertical deflections associated with
the release of a cable force is sufficient in theory to
determine the effective modulus of concrete with the
numerical model. The main limitation of such
determination is due to the significant thermal
variations induced by changes in solar radiation and
rain. The temperature sensors were installed in the
main girders and pylons, and the average thermal
corrections of measured displacements are derived using
long series of vertical displacement and temperature
records. These corrections are found to be around 0.02
m (28% of the maximum) with a coefficient of variation

of about 10%. Through these data the effective modulus of concrete was found to be 45000 MPa.

Conclusions

The ambient vibration measurements performed before initiation of cable replacement have shown to yield valuable data to control the repair process. Acquisition of vibration records was performed in just few hours of a cloudy day, and thus was largely independent of thermal variations. This is in sharp contrast with the static deflection measurements that require a series of observations for two months before meaningful results were obtained.

The effective modulus of elasticity of concrete applicable to the numerical model was found to be 42500 MPa, while the static deflections yielded 45000 MPa. Calculations to define the schedule for profile correction were performed on the basis of the effective modulus of concrete, of the numerical model and of the measured cable forces.

The results of the ambient vibration measurements provided an opportunity to calibrate the numerical model of the bridge, which was also used to control the response of the structure to unusual loading conditions arising in the process of cable replacement.

References

[1] Felber, A., and Cantieni, R., "Advances in Ambient Vibration Testing: Ganter Bridge, Switzerland", Journal of the International Association for Bridge and Structural Engineering (IABSE), Vol.6, No.3, August 1996, pp. 187-190.

[2] Cantieni, R., "Updating if Analytical Models of Existing Large Structures Based on Modal Testing," RECENT ADVANCES IN BRIDGE ENGINEERING: Evaluation, Management and Repair, Edit. J.R. Casas, F.W. Klaiber and A. R. Marí, Proceedings of the US-Europe Workshop on Bridge Engineering, Technical University of Catalonia and Iowa State University, Barcelona, July, 1996.

[3] Ceballos, M. A. et al., "Experimental and Numerical Determination of the Dynamic Properties of the Reactor Building of Atucha II NPP", SMIRT13 Post Conference Seminar 16: Seismic Evaluation of Existing Nuclear Facilities, Iguazú, Argentina, August, 1995, pp. 311-327.

Acknowledgements

This work was done with support from the Research Council of the Province of Córdoba, Argentina(CONICOR).

Behavior of Stainless Steels in Concrete

Pietro Pedeferri, Luca Bertolini, Fabio Bolzoni,
Tommaso Pastore[1]

Abstract

The corrosion behavior of stainless steels in concrete is described. The studies carried out on this
matter are illustrated. The importance of pitting
potential and critical chloride content is underlined.
The behavior of different stainless steels (austenitic
AISI 304 and 316, super-austenitic 254 SMO, duplex
23Cr4Ni, martensitic AISI 410, ferritic 430 and carbon
steel for comparison purpose) in "sound" or carbonated
concrete, with or without chlorides, at low or high
temperature, when stainless steel is coupled or not
with usual steel reinforcement is illustrated. A
comparison between the use of stainless steel and the
main types of additional protection is given.

Introduction

Stainless steels can be used for parts of new
structures subjected to chloride penetration under very
aggressive environmental conditions (e.g. for skin
reinforcement) and for important constructions with a
very long service life. They can also be useful in the
repair of reinforced concrete structures where low
quality or low thickness concrete cover cannot be
avoided.

The stainless steels proposed for concrete reinforcement can be divided, according to their structure,
in three groups: austenitic, ferritic, and austeno-
ferritic. Austenitic stainless steels contain 17-19%Cr

---

[1]Dipartimento di Chimica Fisica Applicata, Politecnico di
Milano, via Mancinelli 7, 20131 Milano, Italy.

and 8-13%Ni (AISI 304), possibly with 2-3%Mo (AISI 316); ferritic ones have 12-17%Cr; austeno-ferritic contain 22-26%Cr, 4-8%Ni and possibly 1-5%Mo (Table 1). A suitable material selection among the wide range of available stainless steels can be made in relation to both environmental aggressiveness and other requirements. However, so far, mainly austenitic stainless steels types AISI 304 and 316 have been used in concrete constructions.

Corrosion behaviour. In environments with neutral or alkaline pH, namely both in carbonated and alkaline concrete, stainless steels are passive, thus they do not suffer general corrosion, but in particular circumstances they can be susceptible to localized corrosion.

Theoretically different forms of corrosion can take place: intergranular, stress corrosion cracking and pitting corrosion. In practice, only the last type of attack is a real problem. Intergranular corrosion is now avoided with steels specially alloyed or with a controlled (low) carbon content, and stress corrosion cannot take place in the normal service conditions but only in combinations of high temperature, carbonated concrete, and high chloride content which are not encountered in real constructions.

Hence, the dangerous form of corrosion for stainless steels in concrete is pitting corrosion. This type of attack initiates when the potential of steel ("corrosion potential") is more positive than a potential value which depends on steel composition and structure and environment, called "pitting potential".

Table 1   Composition and minimum yield strength of some stainless steels used as reinforcement in concrete structures.

TYPE	Structure	Schematic composition	0.2% Yield strength (MPa)			
			(1)	(2)	(3)	(4)
405	ferritic	X2Cr11			350	
304	austenitic	X5CrNi18-10	200	550	460	460
316	austenitic	X6CrNiMo17-12-2	200	550	460	460
22Cr5Ni	aust.-ferr.	X2CrNiMoN22-5-3		800	700	600

(1) fully softened (round plain bar); (2) drawn (ϕ3-12 mm); (3) hot rolled (ϕ8-20 mm); (4) hot rolled (ϕ25-40 mm).

Since pitting potential becomes more negative as chloride concentration increases, in reinforced concrete, a critical chloride concentration for the initiation of localized corrosion can be reached at the reinforcement surface, depending on the steel potential.

This type of attack is the same localized corrosion process that occurs on carbon steel reinforcements in alkaline concrete contaminated by chlorides, although stainless steels can withstand much higher critical chloride contents. Nevertheless, if corrosion initiates, pitting can penetrate even on stainless steels with rates higher than 1 mm/year.

The composition of stainless steels influences their resistance to pitting corrosion. For instance, in acidic and neutral environments, the corrosion resistance of austenitic stainless steels is dependent on the "pitting resistance equivalent", defined as:

$$PRE = \%Cr + 3.3\%Mo + 16 \div 30\%N.$$

Other properties. Besides the corrosionistic ones, many other properties have to be considered in order to use stainless steels as reinforcement, such as mechanical properties, workability, weldability, thermal conductivity and thermal expansion. Also the production processes, cold or warm working, which enable high yield strength to be reached (table 1) are important in defining the characteristics of reinforcement (Nürnberger, 1996). Products are available as bars, wires and meshes.

The cost of a stainless steel depends mainly on the chemical composition: the higher the alloy content, the higher the cost. In 1995 in Europe, ferritic and austenitic stainless steels bars had an average price about respectively 7 and 12 times higher than the price of carbon steel bars (Nurnberger, 1996). Since then the prices of stainless steel reinforcements have slightly reduced. Further reductions are expected, owing to new developments in production and processes techniques (Abbot, 1997; Valbruna, 1997).

However, as far as the cost of stainless steels is concerned, not only the price of the material itself, but all the life cycle costs of a structure should be considered. Doing so stainless steels can be a cost effective alternative to traditional reinforcements mainly in very aggressive environments or for very long service life. To make the best use of the available materials it is important to select the steel which is adequate at the lowest cost: depending on the aggressiveness of the environment, a proper type of stainless steel has to be selected. Furthermore, often stainless steels can be used in conjunction with the usual carbon steel reinforcement (for example only in the more critical areas or in the external layers of reinforcement).

Studies on the Corrosion Behavior of Stainless Steel Reinforcements

The first studies on the behavior of stainless steel reinforcements in chloride contaminated concrete date back to the seventies. They were carried out by the Building Research Establishment in the United Kingdom (Browne, 1975; Treadway, 1989) on concrete prisms of medium and high porosity concrete (w/c = 0.6 and 0.75) mixed with chloride contents from 0 to 3.2% (by cement weight) and cover thickness of 10 mm and 20 mm. After 10 years of exposure, austenitic stainless steels (AISI 304 and AISI 316) did not suffer corrosion, even with the highest chloride content. Ferritic stainless steels (AISI 405 and 430, with 13% and 17% of chromium respectively) showed a good behavior up to 1% chloride.

At the end of the eighties, Sorensen (1990) found, with potentiostatic tests at +200 mV vs SCE, critical chloride contents higher than 5% and 8% respectively for AISI 304 and 316, which were both reduced to 3.5% in the presence of a welding scale on the steel surface.

At the beginning of the ninthies, Callaghan (1993) with the results of accelerated marine exposure tests, showed that a ferritic stainless steel with 12% of chromium might be the best choice in moderately aggressive environments, where galvanized rebar is not sufficiently resistant and the high resistance of the more expensive austenitic stainless steels is not necessary.

In the same years (Pastore, 1991) the use of a duplex austenitic-ferritic stainless steel (23Cr4Ni) was proposed. This steel showed a localized corrosion resistance similar to AISI 304 and AISI 316 austenitic stainless steels. In fact, corrosion monitoring of reinforcements embedded in chloride contaminated concrete evidenced fully passive conditions after 18 months of testing in concrete with chloride contents up to 3% by cement weight.

Nürnberger (1993) confirmed that the austenitic stainless steels type AISI 304 and AISI 316 do not suffer pitting attack in alkaline or carbonated concrete containing up to 2.5% chlorides (with respect to cement) even in the welded condition. He also showed that 11%Cr ferritic stainless steels (either welded or not) do not suffer pitting in alkaline concrete containing up to 2.5% Cl⁻ but are susceptible to a shallow pitting attack in carbonated concrete with 1% Cl⁻ and a severe pitting attack at 2.5% Cl⁻.

Pitting Potential and Critical Chloride Content

An extensive study has been carried out during the last years at Politecnico di Milano on several types of stainless steels in concrete and in solutions simulating the concrete pore liquid, in order to evaluate the corrosion conditions in concrete exposed to various environments (details on experimental and results are presented in Bertolini, 1996).

Austenitic stainless steels AISI 304 and 316, duplex stainless steel 23Cr4Ni, martensitic AISI 410 with 13% Cr, ferritic AISI 430 with 17% Cr, high alloyed austenitic stainless steel 254 SMO with 20% Cr, 18% Ni, 6% Mo, 0,2%N and carbon steel (for comparison) have been considered.

Potentiodynamic tests. Pitting potentials were evaluated through potentiodynamic tests in saturated $Ca(OH)_2$ solutions at temperatures of 20°C and 40°C with chloride content up to 8%. Figure 1a shows that pitting potential for carbon steel at 20°C evidences a reduction of about 750 mV/decade of chloride concentration, starting from potentials typical for oxygen evolution (slightly above 600 mV vs SCE, as obtained in tests without chlorides) to a value of about -400 mV in a solution with 3% chloride. For concentrations higher than 3% no significant change is observed. Conversely, pitting potentials of all the stainless steels are higher than 300 mV vs SCE, even in the presence of high chloride concentrations.

At 40°C, a general reduction of pitting potentials is observed (Fig. 1b) except for 254 SMO. Pitting potential of carbon steel is shifted approximately 200 mV towards more negative values, compared to measurements at 20°C. A remarkable shift is observed on AISI 410 stainless steel.

Potentiostatic tests. The variation of the critical chloride content for pitting corrosion initiation as a function of pH has been studied with potentiostatic polarization tests in solutions with pH ranging from

7.5 to 13.9 where the steel has been polarized at 200
mV vs SCE and chlorides where progressively added (the
concentration has been increased by 0.5% Cl⁻ every two
days until corrosion initiated).

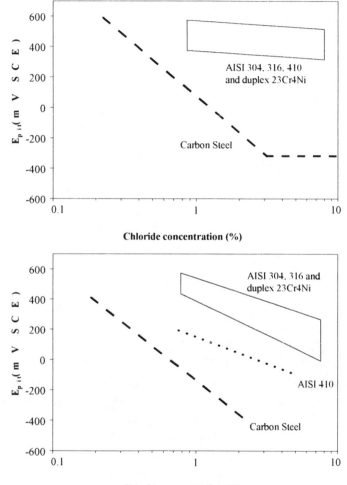

a)

b)

Fig. 1 Influence of chloride concentration on pitting
potential of steel. Based on results of potentiodynamic tests in saturated Ca(OH)₂ solution
at room temperature (20°C, a) and 40°C (b).

At 20°C in saturated Ca(OH)$_2$ solution (pH 12.6) corrosion initiated on low-chromium AISI 410 stainless steel at a concentration of 2% Cl$^-$. Concentrations higher than 5% were reached on austenitic stainless steels, and no localized corrosion attacks were observed on 23Cr4Ni and on 254 SMO stainless steel up to 10% Cl$^-$.

A reduction of critical chloride content to 4.5% for austenitic steels and 3.5% for duplex ones was found during tests at 40°C. No reduction was found for 254 SMO.

Beneficial effect of alkalinity on chloride induced localized corrosion was observed both on carbon steel and stainless steels. All stainless steels had critical chloride content exceeding 10% in the solution with 13.9 pH, both at room temperature and at 40°C. Localized attacks initiated during tests performed at 40°C only on AISI 410 stainless steel at 7% Cl$^-$ and AISI 304 at 10% Cl$^-$.

Test in solutions with nearly neutral pH showed that stainless steels, although still passive, have a lower resistance to chloride induced corrosion, especially regarding the steels with low chromium content.

Tests in concrete prisms. The results of tests carried out in solutions have been confirmed also for steel embedded in concrete. Tests have been carried out in prisms cast with 350 kg/m^3 of portland cement, w/c = 0.55, 1900 kg/m^3 of aggregate and chloride concentrations up to 6%, added as CaCl$_2$ to the mix water.

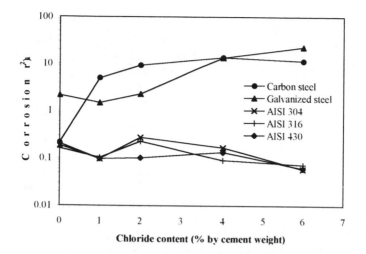

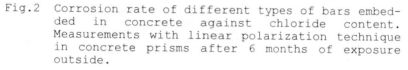

Fig.2 Corrosion rate of different types of bars embedded in concrete against chloride content.
Measurements with linear polarization technique
in concrete prisms after 6 months of exposure
outside.

Each prism contains bars of the different types of
stainless steel; also carbon steel and galvanized steel
are present. Specimens have been exposed outside and
corrosion rate has been regularly measured by means of
the polarization resistance technique.

Fig. 2 reports the corrosion rates measured after
six months of exposure. The austenitic (AISI 304 and
316) and ferritic (AISI 430) stainless steels are in
passive condition in concrete with total chloride
contents up to 6% by cement weight. Corrosion rate is
lower than 0.2 mA/m^2(i.e. lower than about 0.2
µm/year). Carbon steel has a negligible corrosion rate
(<1 mA/m^2) only in the absence of chlorides;
conversely, even a chloride content of 1% is enough to
bring about the corrosion attack. As far as galvanized
bars are concerned, corrosion rate ranges between 1 and
2 mA/m^2 (i.e. still negligible values) in chloride free
concrete and in concrete with chloride contents up to
2%. However the corrosion rate becomes higher than 10
mA/m^2 for chloride contents above 2% of cement weight.

Stainless Steels Behavior in Practical Situations

On the basis of the results here discussed and of previous experience, a schematic representation of the applicability fields of the different types of stainless steels have been drawn in Fig. 3. It shows the conditions where it is safe the use of the different steels as a function of pH and chloride content of concrete (expressed by weight of cement) at 20°C (a) and 40°C (b).

It is evident that carbon steel is susceptible to localized corrosion for low chloride contents even in alkaline concrete. When pH of concrete pore liquid decreases or chloride content increases, stainless steels can be utilized. In fact, in carbonated concrete pore solution pH falls below 9, and carbon steel does not remain in passive state. Low corrosion resistant stainless steels, namely with low chromium content (13%Cr), could be used in this conditions, provided a relatively low chloride content (not exceeding roughly 0.5%) is expected, especially in warm climates (Fig. 2). In alkaline environments, this steel might be useful for chloride contents up to 2% and only for higher chloride concentration (up to 4.5-5%) 18Cr8Ni austenitic stainless steels are required.

The behavior of stainless steels in different practical situations can be summarized as follows.

"Sound" concrete. In sound concrete, i.e. with a pH of 12.5-14 and without chlorides, the film of chrome oxide which forms on the surface of a reinforcement made of any type of stainless steel is as stable and protective as the iron oxide film which covers the normal carbon steel reinforcement. If the characteristics of concrete with regard to reinforcement passivity remain unchanged for the whole service life of the structure, there is no reason, from the point of view of durability, to give preference to a stainless steel reinforcement, which is costlier than normal carbon steel.

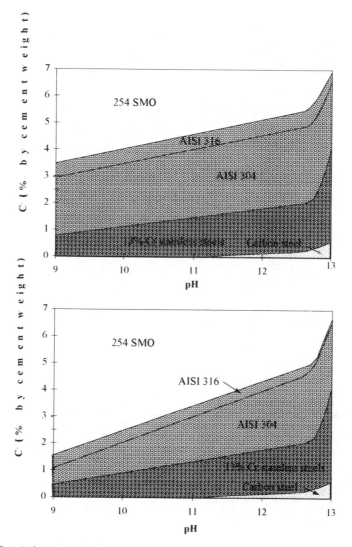

a)

b)

Fig.3 Schematic representation of fields of applicability of different steels at 20°C (a) and 40°C (b).

However, there may be other reasons, connected with the special magnetic properties or the low thermal conductivity of austenitic stainless steel. Or, for example as in the case of anchorages, because the reinforcement is only partly covered by the concrete.

Carbonated concrete. As opposed to the film which forms
on carbon steels (or low alloyed steels) which is
protective only against alkaline environments, the one
which covers stainless steels is protective also in
neutral and slightly acid environments.

Consequently, in carbonated concrete reinforcement made
of stainless steel, with any composition or structure,
works perfectly well, whilst the normal low-alloyed
steel will corrode.

Concrete containing chlorides. As far as concrete contaminated by chlorides is concerned, we have seen that
corrosion of steel reinforcement will not take place if
the ion content falls below a certain threshold, the
critical chloride content.

It has been shown that this threshold depends on
chemical composition and structure of the steel, and
the pH value of the concrete. It also depends on the
potential of the steel, which is related to the
availability of oxygen at the steel surface and,
consequently on whether the structure is exposed to the
atmosphere, is immersed, or is buried. In the most
critical and common case, that of exposure to the
atmosphere, the threshold for normal reinforcement is
about 0.4-1.0% by cement weight.

    For pickled and passivated AISI 304 and 316 steels
it rises respectively to 5 and 8%. However, in cases of
poor surface finishing, the values for the two
stainless steels are reduced and approach each other,
even falling down to 3.5% when oxides produced by
welding are not removed from the surface of the
reinforcement. In this case the behavior of AISI 304
stainless steels and the one of AISI 316 are similar.

    The austeno-ferritic stainless steels behave
similarly to austenitic ones. The critical chloride
content for ferritic steels with 13% chromium is about
2%. No critical chloride content has been found for the
superaustenitic stainless steel 254 SMO.

    All these values refer to temperatures of about
20°C (i.e. those of temperate climates). For
temperatures typical of tropical regions (30°C-40°C),
the values of critical chloride content are slightly
reduced.

Table 2 Contamination of support slabs for bridges, defined by the US Federal Highways Administration.

Grade	kg of chloride on m^3 of concrete	Chloride content (% by cement weight)
Low	0.0 - 2.4	0 - 0.8
Medium	2.4 - 4.8	0.8 - 1.6
High	4.8 - 6.0	1.6 - 2.0
Very High	6.0 - 9.0	2.0 - 3.0

Following the classification of aggressive chloride containing environments given by the US Federal Highway Administration (Table 2), it can be stated that:

- carbon steels (with a critical content 0.4-1% chlorides) can be attacked even when contamination is low,

- austenitic stainless steels, type 304 and 316 (with critical chloride content at least 3.5% even if coated with welding oxides) do not suffer corrosion even in very severe conditions of contamination and temperature,

- ferritic stainless steels, 12-13% Cr (critical content about 1.5-2%) can be used where a medium or low contamination is expected.

Chlorides and carbonation. If concrete, in addition to being contaminated by chlorides, is also carbonated, as may occur on rare occasions (for example, in some highway tunnels), corrosiveness is increased. In fact, as the pH drops, so does the critical chloride content (Fig. 3). In carbonated concrete (pH 9) this content is roughly one third that of alkaline concrete. In carbonated concrete containig chlorides AISI 316 steel behaves better than the AISI 304 one. In selecting the stainless steel, the same criteria apply as for steels partly immersed in concrete.

Steel partially immersed in concrete. At times, the surface of the reinforcement or anchorage of stainless steel is not completely immersed in concrete but is partially exposed to the atmosphere. The choice of steel in such cases is made following the same criteria

as for aqueous solutions with a pH near to neutral. In
practice, steels with increasing contents of chrome and
molybdenum and possibly nitrogen (i.e. going from
ferritic steel to AISI 304, to AISI 316 and possibly to
even higher alloyed steels) are selected as the
chloride content and temperature increase and pH
decreases.

Joining stainless steel and carbon steel. Very often
the use of stainless steel for reinforcement is limited
to the surface of the structure or to the more critical
parts (joints, supports, anchors, or in areas where
corrosivity is highest). As a consequence, galvanic
coupling occurs when stainless steel is connected or
welded to carbon steel reinforcement. Such coupling
does not work if the carbon steel reinforcement is
passive, because in this case its corrosion potential
is comparable to that of the stainless steel. Carbon
steel hence remains passive even after the connection
to stainless steel. On the other hand, coupling works
if carbon steel is allready corroding. In this case
stainless steels bhaves respect to corroding steel as
passive carbon steel does. Nevertheless, the increase
of corrosion rate becomes appreciable only in damp
concrete of low electrical resistivity and for very low
ratios of surface area of carbon steel to stainless
steel, which act respectively as anode and cathode. In
practice, the first condition occurs only in the case
of concrete highly contaminated with chlorides and
therefore already in a condition unsuitable for carbon
steel reinforcement, and the second rarely happens.

Comparison between the Main Types of Additional
Protection

    In Fig. 4 (Pedeferri, 1996) a comparison between
the main types of protection techniques for reinforced
concrete structures in salt laden environments is
reported.

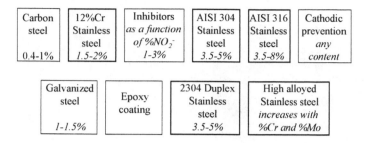

Chloride content (% by cement weight)

Fig.4 Schematization of the maximum chloride content
       at the bars surface in non carbonated concrete
       for different types of stainless steel and other
       additional protection systems.

For each type of stainless steel or technique,
indicative values of the maximum chloride content (% by
weight with respect to cement) which can be reached at
the surface of the reinforcement without causing corrosion are given.

No indication of percentage is given for epoxy
coating of the reinforcement, since there is no
unambiguous documentation available regarding the
maximum chloride content that this reinforcement can
resist.

Acknowledgments

The authors are grateful to Acciaierie Valbruna SpA,
Vicenza Italy, for supporting this work.

References

C.J.Abbot, "The Total Solution for Extended Design Life
in Corrosion Free Concrete Structures", SSR,
Chessington, 1997.

L.Bertolini, F.Bolzoni, T.Pastore, P.Pedeferri
"Stainless Steel Behaviour in Simulated Concrete Pore
Solution", British Corrosion Journal, 1996, Vol.31, 3,
218-222.

B.L.Brown, D.Harrop, and K.W.J.Treadaway: "Corrosion

testing of steels for reinforced concrete, Building
Research Establishment, Garston, 1978, 45/78.

G.Callaghan: *"The Use of 3Cr12 as reinforcing in concrete"* Corr. Sci., 1993, **35**, 1535-1541.

U.Nurnberger, W.Beul, and G.Onuseit: *"Corrosion Behaviour Of Welded Stainless Reinforced Steels In Concrete"* Otto-Graf-Journal, 1993, (4), 225-259.

U.Nürnberger (ed.): *"Stainless Steels in Concrete - State of the art report"*, European Federation of Corrosion, Publication N.18, The Institute of Materials, London, 1996.

T.Pastore, P.Pedeferri, L.Bertolini, F.Bolzoni, and A.Cigada: *"Electrochemical Study on the use of stainless steels in concrete"* Proc. Int. Conf. 'Duplex Stainless Steel 1991', Beaune, France, 1991, Vol. 2, 905-913.

P.Pedeferri, L.Bertolini *"Corrosione nel calcestruzzo e negli ambienti naturali"* Ed. Mc Graw-Hill Italia, Milano, 1996.

B.Sorensen, P.B.Jensen, and E.Maahn: *"The corrosion properties of stainless steel reinforcement"*, 'Corrosion of Reinforcement in Concrete' (ed. C.L.Page), 601-605, 1990, London, Elsevier.

K.W.J.Treadaway, R.N.Cox, B.L.Brown: *"Durability of Corrosion resisting steel in concrete"*, Proc. Instn. of Civil Engineers, 1989, Part 1, 305-331.

Valbruna Acciaierie, Vicenza, Italy, private communication.

Structural Damages Caused by Hidden Overloads in a Reinforced and Prestressed Concrete Parking Garage: Diagnosis and Repair Procedures

Mario Paparoni[1]

Abstract

This paper studies the case of a parking structure, with a roof area used for sports activities, which suffered damages due to overloads generated by heavy pavement finishes and by constructions placed on top of the roof slab. Using the information acquired from this case study, we present a panorama of inherent weaknesses frequently present in reinforced and prestressed concrete parking structures, in spite of their apparent load-carrying capabilities.

It is concluded that several factors, many of them going against common sense, can make these structures damage-prone or even dangerous when unintended uses are demanded from them. Reasons for these situations are given and warnings about the possible dangers are mentioned.

Introduction

This paper will not focus in any technologically new or daring aspects of repairs done on a badly damaged complex structure. Instead it will show how widespread beliefs, cultural factors and typological characteristics tend to make certain deceptively simple structures damage-prone. The structures to be examined here are the very frequently built reinforced-prestressed-prefabricated concrete parking structures, in which we combine cast-in-place reinforced concrete members (columns and beams) with prefabricated, prestressed members (beams and slabs).

These structures can be easily damaged when non-intended uses are imposed on them, though these damages are not always spectacular nor do they always lead to fatalities.

There are several recent examples of damages to such structures. In Chicago, at the 1996 American Society of Civil Engineers Convention, several reported examples of seismic damages were presented, showing more than 40 cases of failures in this type of parking structures, all located on the Pacific Coast of

---

[1]Professor of Civil Engineering, Director of the Center for Applied Research - Universidad Metropolitana Apartado 76.819 Caracas 1070, Venezuela. E-mail: mpapar@conicit.ve

the United States. These accidents generated the need to prepare special code provisions. (Fleischmann, 1996; Wood, 1996). The common element in all cases was the use of prefabricated prestressed concrete standard members. Shear walls were the main seismic resistance vertical members, and the failures occurred by disassemblage of the structures.

In years past, the presence of frequent bracket failures in these structures generated a lot of interest, both experimental and theoretical, so that now we have complete design rules incorporated in most design codes. (Park, 1978).

Vertical loading failures are not very common in parking structures, but are not impossible, if unrestricted access of heavy vehicles is permitted. Normally, however, the limited geometrical clearances work as a natural deterrent against overloading. The case we are going to deal with is one of those rare instances where vehicle loading caused absolutely no harm, but unintended hidden loading caused extensive visible damages in the roof of one of these structures (the roof was designed with the same criteria as the parking floor). Precisely this apparent paradox made the diagnosis rather difficult, because the situation contradicted commonsense logic.

Despite the progress   already made in improving the design of these structures, it is still important to draw the attention to the general public that care and prudence to be exercised with these parking structures. The public should know that they often fail because of widespread misconceptions about their overload tolerance capabilities.   Everybody assumes that cars are very heavy objects, which is not true, as the case presented here shows

Parking Structure Referred to in this Study

The structure which gave origin to this work is located in Caracas, Venezuela. It was designed in 1978, has two floors (ground and first level) originally used for parking, and one roof which houses tennis courts and other low occupational density sports facilities, accessible only to people. Three conventional reinforced concrete and brick structures were also built on this roof.   Apparently these structures, not designed by the original project team, were built of entirely as if they were resting on the ground.

That the designers were tennis court building specialists (not engineers) explains this situation. Most, if not all, of their designs were intended for ground level use. They did not refer to the original structural design data for this project, because of the apparently plausible assumption that parking structures can take lots of loads, without the need for further analyses. It is also unlikely that the designers had access to computer generated data about the load capacity of the structure at the time when these structures were designed. Computer generated data relevant to such a project would have appeared as thousands of numbers decipherable only to a trained structural engineer. So, we have a case where a full fledged ground level project was put on top of a supposedly very capable structure.

The horizontal dimensions of this parking structure are approximately 88 meters by 54 meters. There are no joints in the structural frames. There is a single ramp of access for cars and two separate rigid structures with stairs as pedestrian accesses. The parking structure itself and the adjoining staircases are built on top of cast-in-place concrete piles, encased in a relatively shallow fill. The stairs have strong, tubular reinforced concrete central shafts with the stairs attached around them. Individually they constitute rather rigid structural systems, touching, but not intentionally connected, to the main framed structure, causing unsymmetrical horizontal deflections with temperature changes. Typical spans of the framed areas are 8 and 10 meters for the cast-in-place beams and 16 and 18 m for the prefabricated, prestressed bridge beams incorporated into the structure.

Slabs were also entirely prefabricated, prestressed concrete, the typical Pi (TT) shaped kind. The cast-in place columns and beams form a framed, rigidly jointed structure. The prestressed beams have imperfect, partial nodal connections to the framed structure because they rest on concrete brackets and their support plates at both ends are welded to other plates embedded on the upper surfaces of the brackets. There are no shear connections between the prestressed beam webs and the columns. The whole structure was designed for seismic conditions, but assuming perfect nodal continuity for the prefabricated beams. This assumption is certainly debatable, but the project as it is in other aspects (cast in place beams and columns) can be considered acceptable. Though not paradigmatic, because the brackets were designed with old-style criteria, that is with vertical stirrups, and their construction plan detailing was not at all clear in certain vital points. There was apparently only one design for all the bracket types (three geometrical types for a total of 36). A few of the brackets also had smaller dimensions than the indicated ones. This simply means that, at that time, many people did not see brackets as particularly problematic.

Neither the original structure, nor the additional ones built on top could be considered to have serious project, inspection or construction shortcomings, according to two additional successive studies, performed after the first repairs were done on the damaged structure. As a matter of fact, all the structures immediately conveyed, to the non-expert eye, the impression of being visibly strong and without serious shortcomings, except for a marked tendency for developing roof leaks discharging into the parking floor below, and the generalized presence of cracks in many brackets, and shear and flexure cracks in several cast-in-place beams and shear cracks in the prestressed beams with shorter spans. There were no excessive visible vertical deflections and no partial collapses occurred. There was an increase in visible cracking patterns under the constructed areas located on the roof (one office, one bar and one jai-alai court with high perimeter walls).

There were several attempts to fix these problems after the initial cracks became too visible. The first diagnosis, based mainly on the very visible evidence of bracket failures, attributed the cause of the problems to the parasitic shortening of the prestressed beams. These beams had been welded to their support brackets at both ends, causing in them tensile failures. A local prestressing of the brackets with 30 mm Dywidag prestressing bars was performed on the visibly damaged brackets only. This treatment apparently stopped the progression of the bracket cracking, but obviously could not fix the other problems. No other actions were taken after that.

Two studies followed this initial intervention. One concentrated on a very careful description of the damages, with no intention to find the causes of the problem. The second made an ultrasonic and core-drill sampling of the structure, including the roof leveling fill. This study found that vertical overloading existed and recommended the complete substitution of the damaged brackets, together with their column portions. It diagnosed this problem as originating mainly in the thermal stresses induced on the roof by the daily sunshine cycles. This study recommended to unweld the prestressed beam connections and to externally reinforce the damaged beams. No load removal was recommended, basically because of cost, but it recognized its influence. Another study was then commissioned by the client, who considered the recommended simultaneous bracket and column substitutions as very complicated and expensive, and he asked to look for alternatives.

Making use of all three previous studies, and proceeding with reverse thinking, this fourth study, performed by the author, concentrated on the calculation of the real loads applied to the structure, taking as an indicator the maximum shear capacity admitted by the American Concrete Institute as bearable by prestressed concrete beam webs ($2.65\sqrt{f'c}$ in kg/cm^2) at the onset of cracking, to determine the real vertical loads applied to the structure. It was found that the original safety factor margins had been exhausted in the prestressed concrete beams. Consultations with the original designer of the prestressed beams confirmed that a value of $2.12\sqrt{f'c}$ for factored loads had been taken as a design value for vertical loading only, and that the number of stirrups originally incorporated into the beam webs was sufficient to take the original design loads.

All this meant that there were no doubts about the overloaded condition of the roof slab. Another analysis, performed on the damaged cast-in-place beams, confirmed this conclusion. It was then decided that we had to remove most, if not all, the roof overloads, in order not only to preserve the original safety factors, but also for seismic reasons. In fact, an estimated volume of $1023 \, m^3$ of fill, weighing 2558 tons, plus the estimated 1000 tons of weight of the demolished roof constructions, amounting to a total of about 3600 tons of rubble, was removed from the existing structure. Part of that load will be replaced, with a new lightweight fill plus new lightweight constructions, for a total of 1000 estimated tons of new construction.

It is unnecessary to identify the structure. The author has been involved in several cases like this or even worse. In one case, the non-intended loads came from a water-carrying truck serving an office building during a water shortage. In another case, the culprit was an armored car carrying money every day to and from a bank located on the 4th floor of a building. In another, a supermarket truck destroyed several prestressed slabs in a big new development, simply because the original path, provided with suitably reinforced slabs and beams, was inadvertently changed.

In the author's experience, only cranes win over parking structures as frequent victims of their owners. We, as engineers, must, therefore, do our duty to protect these two problem-prone domestic structural species from their masters.

Simplified Diagnosis

• In the case presented here, loads effectively equivalent to 4 or 5 layers of cars, piled on top of each other and fully filling all the available surface, were easily reached. An innocent looking fill, together with some apparently harmless constructions, can easily be put on top or inside any parking lot structure. The

temptation to use these visibly generous and sometimes unprofitable spaces is very strong.
• The consequences of this event were enhanced by a typical characteristic of these structures which is seldom taken into account, that is their relatively low unit weight, which paradoxically, increases the importance of any nonintended load or overload. (CTBUH, 1980; Paparoni, 1991).
• The original design, optimized for vertical loading only in certain critical members, did not leave room for an overload of this kind.
• Only in very limited instances underdesign or undersizing of members was the cause of the troubles.
• People could not understand why, in the apparently heavily loaded area used for parking, there were no problems. This is because the parking floor and the roof are very similar in construction. This made the diagnosis difficult.

General Comments about Structural Concrete Parking Structures

Reinforced or prestressed concrete parking structures are frequently used as part of building developments. Not only they are generally cheaper and more accessible than underground structures, but they also constitute one of the few loopholes in the building ordinances that permit the legal increase of buildable area for a given development. The philosophy behind this ordinance liberality is that parking areas constitute a social asset, because they serve not only the property where they are located, but also the surrounding neighborhood.

In many cases, though, the choice made in designing a parking structure, or more specifically, the choice of covertly overdesigning a parking structure is made for other reasons. One of them is the relatively easy way in which these structures may, quietly and progressively, be converted into offices, shops, workshops, warehouses, etc. This phenomenon is fueled by the low profitability of parking structures, especially when price regulation applies to them. This frequent practice leads people to believe that the original uses of such structures can be changed without serious consequences, and that all parking structures are somehow inherently overdesigned. In the case of the structure we consider here, such path of development was not followed. The roof was designed for recreational activities and the structure was well built and inspected; however, nobody could stop the overloading in time.

Parking Structures as a Cultural Structure Family

Most free standing parking structures have reasonable, regular and repetitive layouts and employ rather large slab and beam spans. Therefore, the use of prefabricated prestressed concrete beams and slabs is very frequent, leading to a kind of structure which is often difficult to design under seismic conditions, mainly because of the complexities of achieving good nodal connections. These structures possess what can be called deceiving simplicity.

From the point of view of vertical loading, the relatively low passing clearances needed from them (to have more floors) make them suitable for passenger cars only. That is, no cargo use is usually included as a design condition, leading to relatively low vehicular design loads. This situation, together with their relatively open character, without design provisions for dead loads originating in division walls, floor finishes or other permanent contents, implies that their vertical design loads are relatively low. This makes them vulnerable to dangerous situations caused by formally non-intended uses, which generally imply higher-than-designed applied loads.

Thus, the following characteristics are almost always inherent in this structural family:

1) They are structures with a lower weight per volume ratio than other conventional structures. (about 2/3 of the typical unit weight of an apartment building). Their use demands the absence of walls, especially finished pavements, ceilings and other normal building components, present in office or apartment buildings. (CTBUH, 1980; Paparoni, 1991).

2) They are designed for live loads which are lower than those people assume they can take. Modern building codes no longer prescribe the usually high design loading of a few years ago, to lower costs and to be consistent with the progressive weight reduction of modern cars. Most of the time, the effectively applied passenger car loading tends to be lower than apartment or office building loading. This is a frequently true, but a hard to believe fact for most people.

3) They tend to have a lot of medium-span isostatic beams included in their framed structural schemes, because they are usually made of prestressed concrete, which makes them efficient. But they may cause parasitic loads if they are put in place without sufficient storage time to let their initial axial shortening dissipate, especially if they are put in place without sufficient support freedom to let them shrink. In this case they may pull the brackets loose.

4) One must make use of reinforced concrete brackets as a logical means of holding the simply supported prestressed beams. Many times the typical beam end plates end up being welded to the brackets on both ends, or tied in some way to the rest of the structure.

5) Temperature variations tend to affect these structures more than others because of their large non-jointed continuous construction, their relatively lower thermal inertia, and in extreme climates, because of their necessarily open character. In tropical climates, the main factor is the daily action of the strong sunlight, which can cause high surface temperatures on the roof areas, normally non-insulated and not thickly paved.

6) Due to the normally large spans demanded, the prestressed concrete beams selected for their construction come off the shelf, from the normal inventory of prestressed concrete bridge beams, which are pretty standard nowadays. This kind of beam, when used for spans smaller than their optimal ones, have shear failures preceding their flexural ones.

7) From the seismic point of view, these structures can be vulnerable, if beam clearance priorities prevail over structural needs (then only shallow beams are possible). If framed, they tend to be incomplete frames, difficult to analyze realistically. If they use shear walls, these shear walls are difficult to locate. They generally have to be small and, because of the typically large, one-way, thin prefabricated slabs (typically Pi-shaped), diaphragm action is precarious. Recent

seismic experiences show that the structure simply disassembles into individual slabs and beams, and the shear walls hardly have a chance to work. They do not even break. (Wood, 1996; Fleischmann, 1996).

All this means that despite of their deceiving geometrical simplicity, parking structures are rather complex from the point of view of their structural behavior. Very frequently they act as real traps in the sense of reacting nastily when their intended uses have changed. This is especially true when these changes are the result of decisions taken by persons who innocently think the structures have   large visible loading capabilities, and overestimate their strength.

The case presented in this paper shows most of these typical characteristics, and its reaction to overloads was no different from the expected response of this structural family.

Specific Causes

The example shown here had the following specific causals which make it a pathological case, but they are fortunately manageable both technically and economically:

1) The invisibility of an overload caused by an apparently harmless   normal concrete leveling fill in a roof intended to make it fit for sporting activities.

2) The design problems related to reinforced concrete brackets, perhaps the most misunderstood of all reinforced concrete members.
3) A non-apparent shear weakness implicit in all bridge beams when their original design spans are made smaller.

4) An oversight in load flows caused by the confusing presence of non-load bearing beams and changes in the slab orientations.

5) Placing ordinary non-lightweight construction on a parking-lot roof.

6) The unexpected influence of the daily sunlight action creating thermal forces.

7) The unshakable faith in computer programs with non-explicit capabilities.

Pre-Diagnostic Condition of the Structure

• The structure stood for almost two decades with visible damages, but without collapsing or showing any visible signals of distress other than visible cracks.
• There was an initial phase of progressive cracking that was controlled by local prestressing of the damaged brackets, which were the most visibly distressed structural members.
• Post-tensioned beams and cast-in-place beams showed visible, but apparently not dangerous cracking patterns, most of them of the shear-type, especially on the large prestressed beams (spans of 16 and 18 m)
• No excessive deflections were detectable under casual visual inspection.
• Careful studies performed in situ revealed that there were no serious construction

faults like concrete understrength or excessive beam deflections, but in these studies comprehensive structural analyses were not performed in depth.
• Cracking patterns were duly recorded, but they did not appear to be extreme patterns, except for the brackets, which showed large open cracks, of various types.
• The overload was not apparent, because there were no obvious ways of ascertaining it visually, and there were no apparently large live loads applied to the structure.
• The additional constructions added onto the top slab did not look too menacing from the point of view of possible overloading, but they were built with rather heavy materials as if they were resting on the ground.
• As usual, the load capacities assumed in the initial project were almost hidden in the quantity of information provided by the original project computer analysis. Normally, this information is of an archival kind, so most probably whoever did the extra work on the roof, did not want, or did not know, how to look for and solve the information riddle.

Shear Weakness of Pre-Stressed Concrete Bridge Beams When Used in Parking Structures

Perhaps the most important lesson learned in this case pertains to the old question of flexural vs. shear strength in prestressed concrete beams. This corresponds to peculiarity number 3 of our list for parking structures. We believe the other causes of weakness can be controlled by common sense, but this one contradicts intuitive knowledge.

Structural engineers are usually trained to verify the implementation of some very specific design rules, under very specific loading. In other words, they check point conditions, not ranges of behavior. The only exception is, perhaps, seismic design, where we have to verify, besides the normal deterministically prescribed loading, some rather unknown loads which are induced in the structure basically by the displacement caused in the structure itself by the seismic actions.

In our case, the most important aspect, besides the dramatic bracket cracking and visible distresses, was the relative absence of any visible flexural cracks or any perceivable deflections of the prestressed concrete beams. Thus, sudden failures may happen without visible warnings.

The root of this problem lies in the way these beams are optimized for their primary intended use, medium sized bridges, with spans of 30 to 35 meters. When we take one of these beams and use it in a parking structure, the spans are reduced to about one half of the original design. Their linear loading capacity goes up, if determined by flexure by a factor of four, but linear loads limited by shear capacities go up only by a factor of two. We can express it this way:

$$[1] \quad W_{\text{flexure}} = 8M/l^2$$

$$[2] \quad W_{\text{shear}} = 2V^*/l$$

Figure 1 shows the relationships between the corresponding values of M at the onset of cracking and V* with $v_u = 2.65\sqrt{f'c}$ for the beams used in the parking structure we are discussing here. Linear loading capacity decreases with the square of the span if governed by flexure, and with the span, if governed by shear. One can see that below 24 m, for these particular beams, shear capacity governs, and we have 16 m and 18 m spans . This explains the observed shear failures which occurred without flexural cracking being visible. The graph compares the nominal flexural capacity, determined by the onset of cracking (0.20 f'c of tension allowed on the beam) with the shear capacity determined by the maximum allowable shear in the web (independent of the amount of stirrups). Specifically for a span of 16 m, shear capacity reaches only 70% of flexural capacity. For an 18 m span, the relation is 76%. This explains why we detected shear failures only, and not flexural cracking, when an overload occurred. Obviously, with normal loading, none of these limits is reached.

The normal practical design approach is to maintain all the original prestressing, which means that the maximum flexural capacity is made use of (independently of the span), and the shear capacity is simply verified for the prescribed design conditions. If the criteria for shear are satisfied, then the beam is considered acceptable. Checking of possible overloads is not normally performed, so this beam, now used in a shorter span, can very well have a shear failure before the onset of flexural cracking. Such failures occur as a result of unintended or unexpected overloads and the beam will not tell very visibly the user that something is wrong. Of course, the effect of the shorter spans of the beams can be compensated for with some local overdimensioning of the beam ends, but this increases the cost of the beams, because it requires a change in the normally heavy steel formworks used to manufacture these beams. For this reason this is seldom done.

Thus, standard bridge beams, when used for shorter than the designed spans, can create unsafe situations in the case of overloads. This is a result that not many people expect.

Types of Reinforcements Chosen for Our Case

Figures 2, 3 and 4 show how the existing brackets were reinforced by encapsulating them, together with their columns, in a reinforced concrete cocoon. This reduced the need to perform precision drilling on the columns, as a steel encapsulation would have demanded. Shear connectors were used between the new covers and the columns. The resulting new brackets are not classical in form. Its polyhedrical shape was chosen both to minimize space intrusion characteristics and the amounts of concrete required.

Figure 5 shows the original reinforcement of the brackets, done with old style procedures: vertical stirrups and undetailed horizontal reinforcements.

Figures 6 and 7 show the sectional shapes of both the beams and the slabs.

Cast-in place beams were reinforced for shear only by putting additional stirrups encased in concrete around their ends.

Acknowledgement

This is a curious case in which the author cannot name the owner of the parking structure studied in this paper, even though he gave full permission to publish the case. This is because there is a strange market law which says that repaired structures loose value, no matter how much stronger and better they may be after installing the reinforcements. Honest owners who repair damages are punished, but those who cover the damages with invisible cosmetic treatments do not suffer such fates. This is what modern neoliberals call the market intelligence.

The author also   thanks INGEMARTIN S.R.L., contractors for the performed repairs, represented in this case by Eng. José Martin and Eng. Pedro Sanglade, for their collaboration.

Universidad Metropolitana allowed the author to write this paper, a non-income activity that many private universities cannot afford.

The case presented comes from the author's private practice.

References

Fleischman, R.B.; Sause R.; Pessiki, S. - 1996
"Seismic Behavior of Precast Parking Structure Diaphragms" American Society of Civil Engineers, Structural Congress XIV, Chicago.

Wood, S.L.; Stanton, J.F.; Hawkins, N.M. - 1996
"Performance of Precast Parking Garages in the Northridge Earthquake. Earthquake Lessons Learned". American Society of Civil Engineers, Structural Congress XIV, Chicago.

Paparoni M., Mario - Feb. 1991
"Dimensionamiento de Edificios Altos de Concreto Armado". SIDETUR - Caracas.

Council on Tall Buildings and Urban Habitat (CTBUH) - 1980
"Tall Buildings Criteria and Loading" Vol. CL.   American Society of Civil Engineers. New York.

Park, Robert; Paulay, Thomas - 1975
"Reinforced Concrete Structures" John Wiley & Sons. NY.

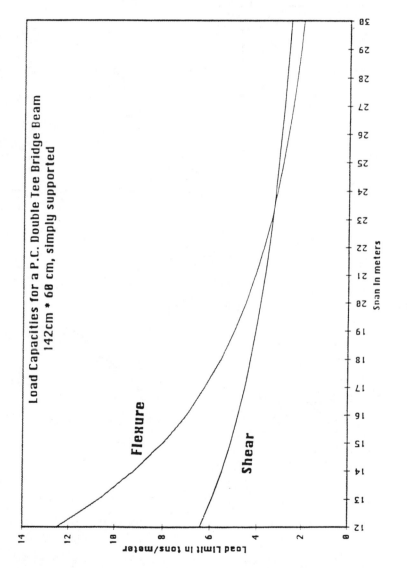

Fig. 1

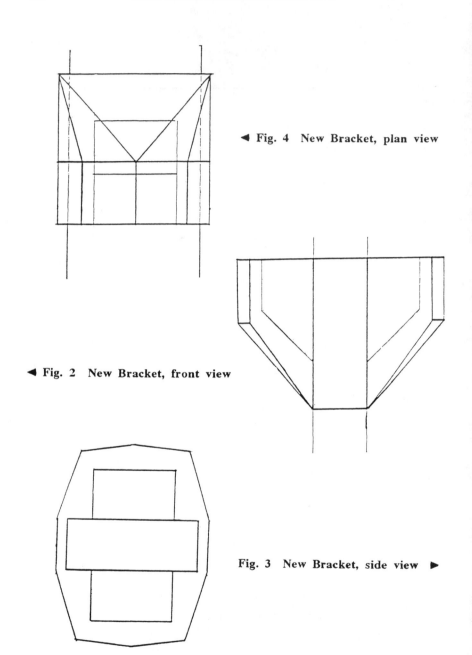

◄ Fig. 4   New Bracket, plan view

◄ Fig. 2   New Bracket, front view

Fig. 3   New Bracket, side view   ►

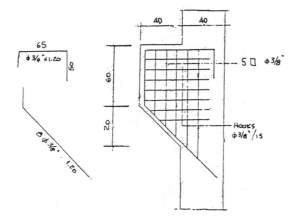

◄ **Fig. 5   Brackets, Original Reinforcement**

Fig. 6   Prestressed Beam Shape ►

◄ **Fig. 7   Prestressed Slabs**

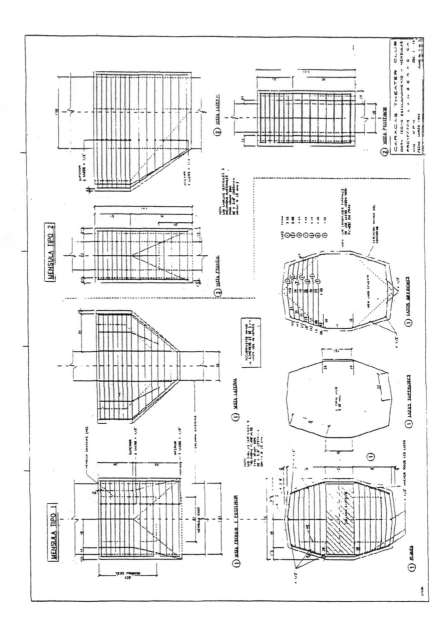

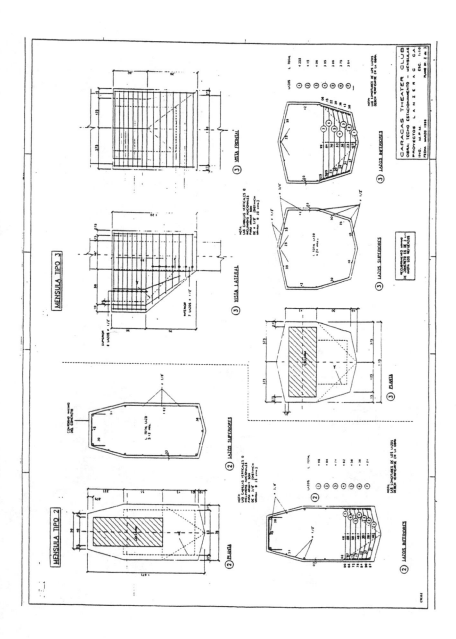

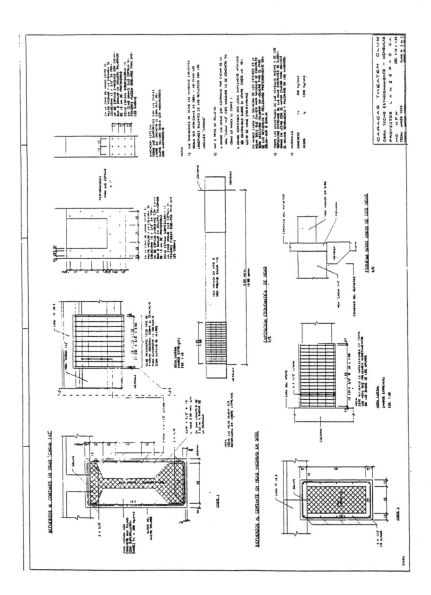

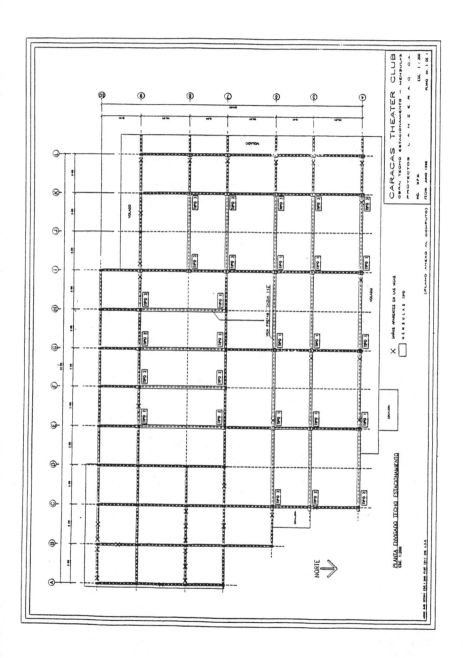

PROTECTION CORROSION MECHANISMS INVOLVED IN REPAIR SYSTEMS WITH SURFACE COATINGS ON REINFORCED CEMENT

E. Pazini[1] and C. Andrade[2]

Abstract

The coating of reinforcement (primer) is usually a component part of a complete system for concrete repair. Howere, few studies have been devoted to evaluate their efficiency in protecting the reinforcement against further corrosion.

In this study, five types of coatings, each with a different protection mechanism, were applied on reinforcing steel bars and embedded in a standar mortar with admixtures. The behaviour of the bars was monitored and compared to a reference specimen. The Polarization Resistance technique was used to measure the corrosion intensity (icoor).

After more than a year of testing, the corrosion trends observed indicate significant differences among the coatings. The results indicate that not all the coatings were equivalent from the corrosion protection point of view.

Introduction

The selection of a repair system for reinforced concrete is influenced by several factors. Among these, one of the most significant factors is the structural condition of the concrete structure component. This being of special

---

[1] Federal University of Goiás, Brazil
[2] Inst. Eduardo Torroja of Construction Sciences, CSIS, Spain

importance in cases were significant loss of cross-section of the reinforcement has accurred.

The basic principles to be considered when repairing concrete have been described by Schiessl (Schiessl, 1992). When the method selected is the traditional patching system, the complete repair system shall include the following (Lambe and Humphrey, 1990):

. Diagnosis and specification of kind of intervention.
. Removal of all damaged concrete back to a sound core.
. Preparation of concrete substrate and reinforcement (cleaning).
. Coating of reinforcing steel.
. Priming of concrete substrate (bond coat).
. Application of repair mortar.
  Application of surface treatment over the entire structure to avoid further carbonation or ingress of chlorides.

The properties and requirements of the coat (Climaco and Regan, 1989; Judge and Lambe, 1987), the repair mortar (Nepomuceno and Andrade, 1992; Paillere et al.) and the surface treatment (Swamy and Tanikawa, 1990; Kamal and Salama, 1990) have been vastly studied and reported. However, very few studies have been conducted to evaluate the performance of coatings on the reinforcement, in spite that this type of material is produced by many manufacturers, recommended by technical documents (Concrete Society, 1984; ACI, 1985), and utilized in large scale repair projects.

In this paper the protective capability, or chloride resistance, of five different types of coatings for reinforcement is discussed. The Polarization Resistance technique was used to monitor the corrosion intensity of the coated rebars. The behaviour of the protected bars is then compared with that of uncoated bars embedded in a specimen of the same mortar type.

Experimental methods

Mortar specimens were fabricated with dimensions of 2 cm x 5.5 cm x 8 cm. The mortar was fabricated with a water-cement (w/c) ratio of 0.5 and a cement-sand (c/s) ratio of 1/3. The cement used was a Rapid Hardening Portland Cement. The chemical composition of the cement used is given in Table 1. In all specimens, 3% CaCl per weight of cement was added to the mixing water in order to develop prompt corrosion activity.

Two corrugated steel bars of 6 mm in diameter and length of 8 cm were embedded in each specimen as duplicate testing electrodes. The chemical composition of the steel is given in Table 2. The steel bars were cleaned as per ASTM G1 (Standard Practice for Preparing, Cleaning, and Evaluating Corrosion Test Specimens) before being embedded in the specimens. This procedure gave the steel bars identical surface grade cleaning before the coating was applied. A fixed area of attack of 5.6 cm^2 was used. This was performed applying adhesive tipe on the bars surface. A graphite rod was embedded between the two steel bars, to serve as counter electrode. Figure 1 shows a detail of the typical test specimens. The figure shows the wire connection to the rebar and the sealing of the external portion of the rebar with a silicone mastic to prevent atmospheric attack on this area.

Table 1. Chemical composition of cement used in reference mortar

Chemical Characteristics	Cement use in reference mortar (%)
P.F.	3,13
R.I.	1,92
SO$_3$	3,08
SiO$_2$	18,32
Al$_2$O$_3$	5,43
Fe$_2$O$_3$	3,28
CaO	61,34
MgO	1,51
Cl	0,02
Na$_2$O	0,15
K$_2$O	1.04

Table 2. Chemical composition of steel electrodes

Element	C	Mn	Si	P	S
Composition (%)	0,17	0,59	0,25	0,02	0,03

Table 3 shows the characteristics of the coatings tested, and of the mortar used for casting all the specimens. All coatings are commercially available and marketed specifically for repair of reinforced concrete. After casting, the specimens were placed in controlled condition chambers with 100% RH and 20 ± 2°C. The exposure conditions of the reference specimen was changed to partial imersion condition after 103 days of testing.

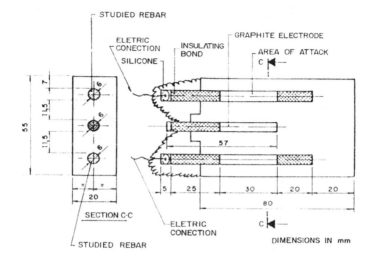

Figure 1. Details of Test Specimen

All the other specimens were maintained under the 100% RH condition until 128 days of testing. After this, all the other specimens also were placed under partial imersion condition. This decision was made in an attempt to accelerate the corrosion process. Figure 2 presents the exposure schedule for all the specimens.

Table 3. Materials tested

COATING OF REBAR	COMPOSITION	No. OF COMPO- NENTS	WET THICKNESS OF THE COATING			TOTAL THICKNESS	pH
			1st COAT	2nd COAT	3th COAT		
refer.	Mortar (1:3:0,5): cement RHCP + standard sand	–	–	–	–	–	13,15
1	Cement+Polymer+ Minerals admixture	1	750μm	650μm	–	1400μm	12,85
2	Cement(Inhibitor+ Acrilic dispersion (Unhibitor)	2	800μm	700μm	–	1500μm	12,11
3	Epoxy	2	350μm	–	–	350μm	–
4	Epoxy/Zinc	1	175μm	175μm	–	350μm	8,48
5	With lead	1	100μm	100μm	100μm	300μm	8,31

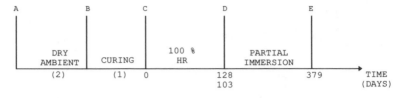

A Cleaning of the rebars   B Coating   C Casting
D Change of conditions exposure   E Finish testing (break of
                                    speciment)

(1) Variable time in function of kind of coating and in function
of the manufactures recommendations;
(2) Variable time defined in function of (1).

Figure 2. Scheme showing the evolution of the test

The corrosion density, icorr, was measured by means of
the   Polarization   Resistance   technique,   where   a
polarization from -10 to + 10 mV around the Ecorr was
applied and the ohmic drop was compensated by means of
the   possitive   feed-back   of   the   potenciostat.   The
intensity changes were determined at a sweep rate of
10mV/min (Andrade et al, 1986). The icorr was calculated
by means of Stern and Geary (1957) equation. The range
between 0,1-0,2 $\mu A/cm^2$ was considered as the frontier
between negligible and significant corrosion (Andrade and
González, 1978). Electrical resistance, Rohm, corrosion
potencial, Ecorr, and polarization resistance, Rp, were
monitored all along the 379 days of testing. The
corrosion potencial, Ecorr, was always measured in
relation to a Satured Calomel Electrode.

Results

The evolution of the variables controlled during the
testing, icorr and Ecorr are shown in Figures 3-9.

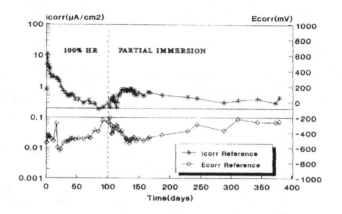

Figure 3. Evolution of icorr and Ecorr values of
reference material

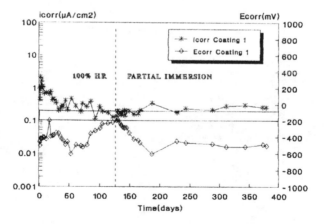

Figure 4. Evolution of icorr and Ecorr values of polymer
modified cementitious material with mineral admixture
primer (Coating 1)

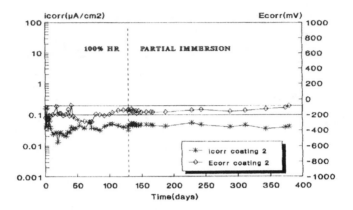

Figure 5. Evolution of icorr and Ecorr values of polymer modified cementitious material with inhibitor primer (Coating 2)

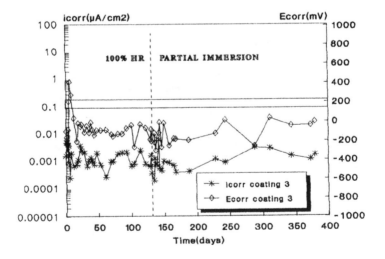

Figure 6. Evolution of icorr and Ecorr values of epoxy resin primer (Coating 3)

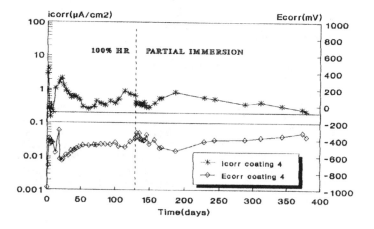

Figure 7. Evolution of icorr and Ecorr values of epoxy zinc rich primer (Coating 4)

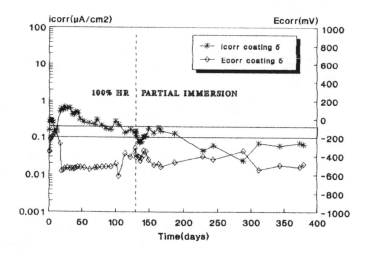

Figure 8. Evolution of icorr and Ecorr values of lead based primer (Coating 5)

The reference specimen as well as coating specimens 1, 4 and 5 exhibited icorr values higher than 0.2 $\mu A/cm^2$. However, it is important to point out that Coating 5 exhibited a decrease of icorr to values as low as 0.05 $\mu A/cm^2$ during the partial immersion phase of the testing period (Figure 8). Coatings 2 and 3 did not exhibited icorr values higher than 0.05 $\mu A/cm^2$. Expecept for Coating 3, the trend of the Ecorr in all specimens was a mirror image of the icorr.

Figure 9 shows the evolution of Rohm during the testing to all coatings. The Coatings 1 and 5 presented Rohm lower than reference. The Coatings 2, 3 and 4 were more resistivities than reference. The low permeability of coating 3 (epoxy resin based) and the high alkalinity inhibitor and polymer admixed in coating 2, may explain the neglible icorr measurements registered on these two cases (icorr lower than 0.05 $\mu A/cm^2$).

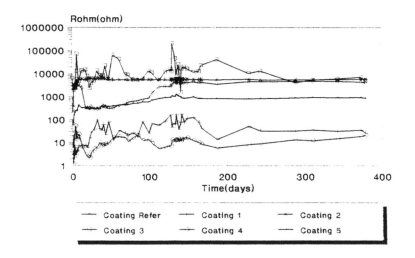

Figure 9. Evolution of Rohm to all coatings tested

Gravimetric weight loss measurements were made and compared to the weight loss calculated from the electrochemical measurements. Figure 10 shows a comparison of gravimetric and electrochemical weight

losses. The diagonal lines represent the error factor of
two suggested by Stern and Weisert (1959).

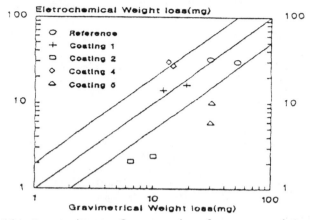

Figure 10. Comparison of corrosion losses as determined
by eletrochemical and gravimetric

Only Coating 4 (rich zinc epoxy) exhibited an
electrochemical weight loss higher than the gravimetric
one. The greater differences between gravimetric and
electrochemical weight loss were registered at icorr
values lower than 0.2 $\mu A/cm^2$ (Coating 2 and 5) due to
the inaccuracy of low weight measurements. Because of the
strong adherence of Coating 3 to the reinforcement steel,
the gravimetric weight loss on this specimen could not be
obtained.

Table 4 shows the percentage of zinc in oxide form
obtained by SEM, measured in three regions of Coating 4.
When dry, the highest quantity of zinc measured was
74.8%, and was registered at the interface between the
coating and the reinforcement.

Table 4.Percentage of zinc in the dry
film mass (SEM) of coating 4

Region studied	ZnO ((%)	SiO_2 (%)	Al_2O_3 (%)
Surface	56,77	31,81	11,42
Center	65,03	25,00	9,97
Interface Conting/steel	74,80	20,57	4,63

Discussion

The are several ways by which coating may protect
reinforcement against corrosion:

a)   Passivation by alkaline materials.
b)   Inhibition by anticorrosive pigments or inhibitors.
c)   Cathodic protection by active admixtures.
d)   Barrier behaviour by decreasing or stopping the
      access of water, oxygen, or corrosive agents to the
      metal surface.

Each protetive mechanism depends on specific chemical and
physical-chemical properties of the coating. Also, in
regard to long term stability, it is important to take
accont that the coating be in contact with a highly
alkaline matrix mortar. A coating may prevent corrosion
by one or more protective mechanism.

Protective mechanism "a" (passivation) appears not to be
effective in the presence of chloride ions due to their
depassivating character. Thus, as Figures 3 and 4 show,
icorr values beyond 0.2 $\mu A/cm^2$ were measured for the
reference mortar (no coating), and Coating 1. Although
high pH levels (12.5 - 13.5) were present, this was not
sufficient to stop the action of the chloride ions.

Protective mechanism "d" (barrier) was clearly identified
in Coatings 2 and 3. These two materials presented a Rohm
value of   6,000 ohms (six times higher than reference
specimen and Coating 1).

It is apparent that Coating 5 could have protected the
reinforcement by corrosion inhibition due to the presence
of lead pigments since it is believed that lead exhibits
inhibitor properties. The mechanism by which lead
pigments act has not yet been fully explained and appears
to differ in a case by case basis.Two explanations of
this mechanism were sound in the literature: a) The first
explanation refers to the properties of the lead powder
which reacts with greasy products on the steel surface
forming a protective metallic soap that repels water
(Grignard and Masson, 1986). b) The second explanation
refers to the property of stifling rusting at its birth
(Fancutt and Hudson, 1957). In addition, a cathodic
protection effect due to the rapid dissolution of lead in
alkaline media can not be forgotten. On this specimen it
was also noticed that the decrease of icorr was
accompanied by an increase in Rohm (Figure 9).

On Coating 4, the protective action of the zinc rich coating may be described as a two stage process (Feliú et al, 1989). A cathodic protection reaction occurs on the first stage. The zinc particles are anodically dissolved and protection of the steel substrate (cathode) is achieved. For this reaction to occur, the zinc particles must be in electrical contact with each other and with the steel base. This is only possible when the zinc particle concentration in the dry coating is very high. In the most favorable case however, Coating 4 presented a 74.8% zinc content (Table 4). The Bristish Standards Institution 4652 (1971) specifies a minimum zinc content of 88% of the dry film mass for this type of coating. The Building Research Station Digest 109 (1969) is more strict, requiring a minimum zinc content of 95% of the dry film mass. Since the maximum dry film zinc contente measured in this specimen was below that specified in the reference literature, it is assumed that the low zinc content was not sufficient to protect the reinforcement as it was expected. The low level of zinc in the coating did not produced polarization of the reinforcing steel to cathodic protection levels. The second stage is a consequence of the first stage and occurs when the corrosion products of the zinc particles fill the pores of the film, resulting in a barrier effect (Feliú et al, 1989). This phenomenon is shown in Figure 7, where the positive shift of Ecorr during testing is accompanied by a decrease of icorr and an increase of Rohm.

Conclusions

The laboratory test method Polarization Resistence may be used to select the more adequate coating to protect the reinforcement against corrosion due to chloride action. The technique was successful in determining the different protective mechanism which acted on the five coatings studied.

There was an acceptable agreement between the electrochemical weight loss obtained by Polarization Resistence and the gravimetrio weight loss measured.

It would not be efficient to repair concrete structures subjected to chloride action using Coating 1. Additional evaluation must be carried out with Coatings 4 and 5 to verify their effectiveness in protecting the reinforcement against corrosion by chloride action. Coatings 2 and 3 presented adequate corrosion control performance in this experiment.

Acknowledgements

The authors are grateful to the Spanish CSICYT for their
finaltial support of this research. Also they thank to
the National Council of Scientific an Technology
Development of Brazil for the grant given to E. Pazini.

References

American Concrete Institute. Corrosion of Metals in
Concrete. ACI/Committee 222R-85, American Concrete
Institute Journal, Proc. 81(1), 1985, pp.3-32.

Andrade, C; Castelo, V.; Alonso, C. and González, J.A.
The Determination of the corrosion Rate of Steel Embedded
Methods - ASM STP 906, ed. V. Chaker, American Sociaty
for Testing and Materials, Philadelphia, 198 , pp. 43-63

Andrade, C. and González, J.A. Quantitative Measurements
of Corrosion Rate of Reinforcing Steels Embedded in
Concrete Using polarization Resistance measurements -
Werkstoffe und Korrosion, vol. 29, 1978, p. 515-519.

British Standards Institution. Specification for metallic
Zinc-Rich Priming Paint. BS 4652, London, 1971,p. 14.

Building Research Station Digest. Zinc-Coated
Reinforcement for Concrete. BRS Digest 109, sept. 1969,
p. 7.

Climaco, J.C. and Regan, P.E. Evaluation of Bond
Strength Bethween Old and New Concrete - International
Conference on Structural Faults & Repair. London, 1989,
8p.

Concrete Society. Repair of Concrete Damaged by
Reinforcement Corrosion - Report n° 26, London, oct. 1984,
31p.

Concrete. ACI Committee 222R-85, American Concrete
Institute Journal, Proc. 81(1), 1985, pp. 3-32.

Fancutt, F. and Hudson, J.C. Protective Painting of
Structural Steel - Chapman & Hall, Londres, 1957, p. 102.

Feliú, S.; Barajas, R.; Bastidas, J.M. and Morcillo, M.
Mechanism of Cathodic Protection of Zinc-Rich Paints by
Eletrochemical Impedance Spectros Copy; Galvanic Stage -

Journal of Coating Technology, Vol. 61, n° 775, aug. 1989, pp. 63-69.

Grignard, R. and Masson, J.C. la Peinture   en   Bâtiment Ed. Eyrolles, Paris, 1986, p. 289.

Judge, A.I. and Lambe, R.W. Selection and Performance of Substrate Priming Systems for Cementitious Repair Mortars.International Conference on Structural Faults & Repair. London, 1987, 21p.

Kamal, M.M. and Salama, A.E. Protection of Reinforced Concrete Elements Against Corrosion by Polymer Coatings Protection of Concrete. E. & F.N. Spon; ed. Dhir, R.K. and Green, J.W., U.K., 1990, pp. 281-286

Lambe, R.W. and Humphrey, M.J. Developments of Materials for Repair and Protection of Reinforced Concrete-Corrosion of Reinforcement in Concrete; SCI and Elselvier Applied Science, ed. C.L. Page, K.W.J. Treadaway and P.B./Bamforth, London, 1990, pp. 479-484.

Nepomuceno, A. and Andrade C. Chloride and Carbonation Resistance of Several Repair Mortars - International Conference on Rehabilitation of Concrete Structures, RILEM/CSIRO/ACRA Melbourne, Australia, 1992, 12p.

Paillere, A.M; Cochet, D. and Serrano, J. Protection Contre la Corrosion des Armatures Apporteé par les Mortiers de Surface a Base Polymeres-Adhesion Between Polymers and Concrete Bonding: Protection and Repair. Chapman and Hall,ed. H.R. Sasse, N. York, pp. 273-288.

Schiessl, P. Repair Strategies for Concrete Strutures Damaged by Steel Corrosion. International Conference on Rehabititation of Concrete Structures, RILEM/CSIRO/ACRA,Melbourne, Australia, 1992, pp. 1-21.

Stern, M. and Geary, A.L. Electrochemical Polarization: a Theoritical Analysis of the Slope of Polarization Curves - Journal Electrochemical Society, 104, 1957, p.56.

Stern, M. and Weisert, E.D. Experimental Observations no the Relation Between Polarization Resistance and Corrosion Rate - Proc. ASTM, Vol. 59, 1959, pp. 1280-1290.

Swamy, R. N. and Tanikawa, S. Surface Coatings to Preserve Concrete Durability - Protection of Concrete. E

& F.N. Spon; ed. Dhir, R.K. and Green, J.W., U.K., 1990,
pp. 149-165.

SUBJECT INDEX

Page number refers to the first page of paper

AUTHOR INDEX

Page number refers to the first page of paper